THE ROYAL SOCIETY OF CHEMISTRY
MONOGRAPHS FOR TEACHERS No. 34

Elementary Organic Stereochemistry and Conformational Analysis

B. A. MARPLES, BSc, PhD, CChem, FRSC
University of Technology, Loughborough

LONDON: THE ROYAL SOCIETY OF CHEMISTRY

Monographs for Teachers

This is another publication in the series of Monographs for Teachers which was launched in 1959 by the Royal Institute of Chemistry. The initial aim of the series was to present concise and authoritative accounts of selected well-defined topics in chemistry for those who teach the subject at GCE Advanced level and above. This scope has now been widened to cover accounts of newer areas of chemistry or of interdisciplinary fields that make use of chemistry. Though intended primarily for teachers of chemistry, the monographs will doubtless be of value also to a wider readership, including students in further and higher education.

First published October 1981

Published by The Royal Society of Chemistry, Burlington House, London W1V 0BN, and distributed by The Royal Society of Chemistry Distribution Centre, Blackhorse Road, Letchworth, Herts SG6 1HN

Printed by Adlard & Son Ltd, Bartholomew Press, Dorking

CONTENTS

1. Introduction
Basic Stereochemical Concepts

Scope of modern stereochemical studies

Stereoisomers are molecules containing the same functional groups and gross structure but they differ in the spatial arrangement of the component atoms. Stereoisomers may be further classified as configurational or conformational and, generally speaking, the former are interconverted by the *breakage* and reformation of a bond (single or double) while the latter are interconverted by *rotation* around a single bond. This distinction is sometimes found to be artificial, as we shall see, and some stereoisomers may be described by

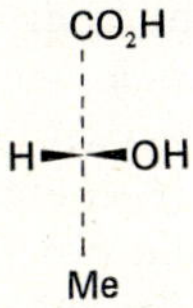
1

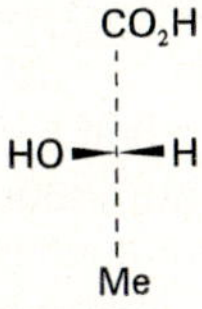
2

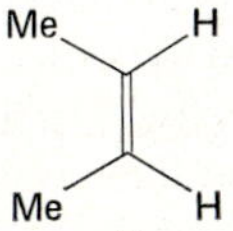
3

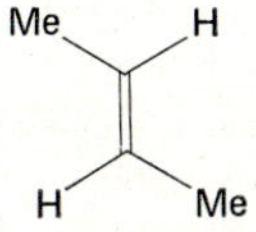
4

either term. However, the generalisation is useful. Thus, *laevo*[†]-lactic acid (2-hydroxypropanoic acid) *1** is the configurational isomer (enantiomer) of *dextro*[†]-lactic acid *2* and *cis*[§]-but-2-ene *3* is the configurational isomer (diastereoisomer) of *trans*[§]-but-2-ene *4*.

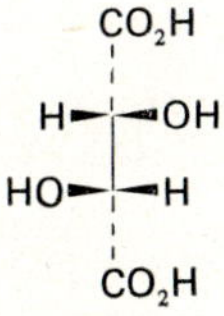
5

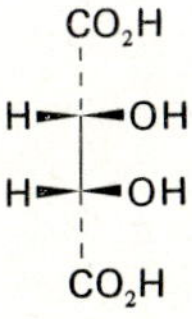
6

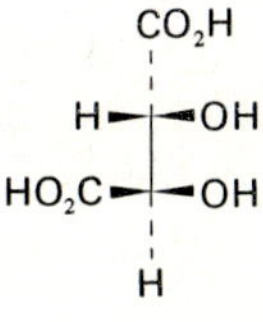
7

A similar diastereomeric relationship exists between *dextro*-tartaric acid (2,3-dihydroxybutanedioic acid) *5* and *meso*[‡]-tartaric

* A bond shown as a wedge indicates that the group or atom is above the plane of the paper; a dashed bond indicates that the group is below the plane of the paper.

† *laevo*- or *dextro*- indicate that solutions of these molecules rotate the plane of plane-polarised light to the left or right respectively.

§ In alkenes of the general types Cab:Cac and Cab:Cab the *cis*-alkene is that in which the two groups a are on the same side of the double bond and the *trans*-alkene is that in which the groups a are on opposite sides of the double bond.

‡ A *meso*-compound is an optically inactive diastereoisomer of an optically active compound.

acid *6* whereas the *dextro*-tartaric acid *7* differs from *5* in its conformation* since C-3 has been rotated through 120° relative to C-2.

Traditionally, stereochemistry was the study of stereoisomerism of the configurational type (enantiomerism and diastereoisomerism) and an understanding of this type of isomerism was initiated by the independent suggestions of van't Hoff and Le Bel (1874) that the valencies of carbon are tetrahedral. Only after this was it possible to understand the structural differences in *dextro*-, *laevo*-, and *meso*-tartaric acids which had been remarkably separated earlier by Pasteur. It was considerably later before the importance was realised of the influence of preferred conformation on physical and chemical properties (conformational analysis). Early indications of this in acyclic compounds came through the observation of stereoisomerism in biphenyls (1926) and Böeseken's study (1928) of the effect of diastereomeric butan-2,3-diols on the conductivity of boric acid. Shortly after this (1930) came the independent suggestions by Weissberger and Wolf from a study of the dipole moments of stilbene dichlorides that rotation around carbon–carbon single bonds was not entirely free. The importance of conformational changes in cyclic compounds was initiated by the observation by Sasche (1892) that the cyclohexane ring could be puckered and maintain the tetrahedral bond angles for carbon thereby avoiding any bond angle strain (Baeyer 1885). Mohr (1918) indicated that the now familiar chair conformers may be interconverted by rotation around carbon–carbon single bonds [*8*⇌*9*], and Kohlrausch (1936) recognised the existence of two distinct types of bond (axial and

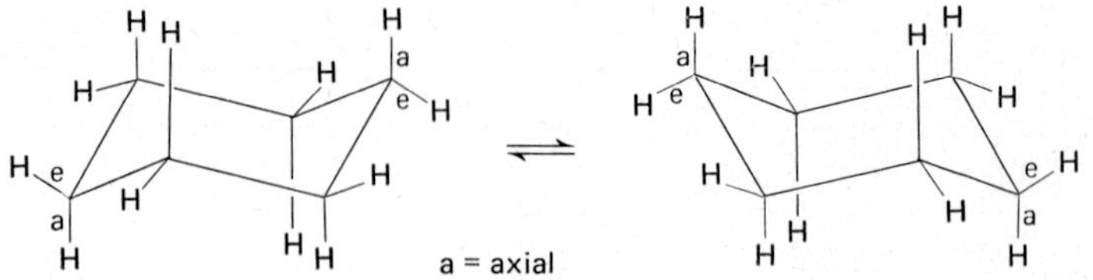

8 *9*

equatorial) in these chair conformations. However, the most significant contributions to our understanding of cyclohexane conformational analysis came from Hassel and Barton for which they were jointly awarded the Nobel Prize for Chemistry in 1969.

It is clear that modern stereochemistry includes a study of conformational analysis as well as the traditional study of enantiomers and diastereoisomers (static stereoisomerism). The separation in

* It is not implied that either of these conformations is the minimum energy conformation.

the title of this monograph of these dynamic aspects of stereochemistry is made to emphasise their importance, and to indicate that the greater part of the text will be devoted to these aspects.

Chirality, enantiomers and diastereoisomers

A molecule or an object is chiral if it cannot be superimposed on its mirror image. The term chiral was derived (Thompson 1884) from *cheir*, the Greek word for hand, a familiar chiral object. A molecule and its non-superimposable mirror image are known as **enantiomers**, and such molecules are devoid of a centre of symmetry, a plane of symmetry and an improper rotation axis of symmetry. In the absence of these symmetry elements the molecule is termed dissymmetric and if, in addition, the molecule lacks a proper rotation axis (other than C_1), the molecule is asymmetric. It is helpful to consider these four symmetry elements in further detail.

A molecule possesses a **proper rotation axis** of order n (C_n) if rotation about the axis of $(^{360}/_n)^\circ$ produces a new structure indistinguishable from the original. All molecules possess an infinite number of trivial C_1 axes because rotation by 360° about any axis restores the original structure. Also, linear molecules such as acetylene or nitrogen possess a C_∞ axis which is coincident with the linear molecular axis. Benzene has a C_6 axis perpendicular to the plane of the ring and passing through its centre.

A molecule possesses a **centre of symmetry** (i) when a line drawn from each atom through the centre encounters an identical atom equidistant from the centre. This is exemplified by α-truxillic acid *10* which, because of the presence of this symmetry element, is achiral.

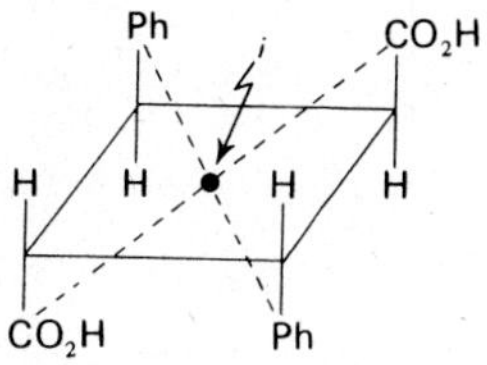

10

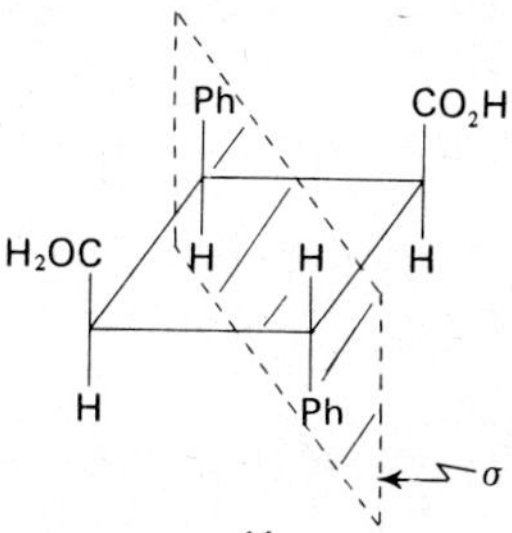

11

A **plane of symmetry** (σ) is a plane passing through the molecule such that all the atoms on one side of the plane are reflected through the plane as though it were a mirror. Such a plane of symmetry is present in *meso*-tartaric acid *6*, in which it bisects the 2,3-bond, and in γ-truxillic acid *11*, in which it bisects the cyclobutane ring diagonally and contains the two phenyl groups. Accordingly, both these compounds are achiral.

A molecule possesses an **improper rotation axis** of order n (S_n) if rotation about the axis by $(^{360}/n)°$ followed by reflection through a plane perpendicular to this axis produces an equivalent structure. For example, α-truxillic acid *10* has an S_2 axis which passes through the centre of and perpendicular to the cyclobutane ring. As indicated in *Fig. 1* reflection of the structure produced by rotation through 180° through a plane perpendicular to this axis produces an equivalent structure.

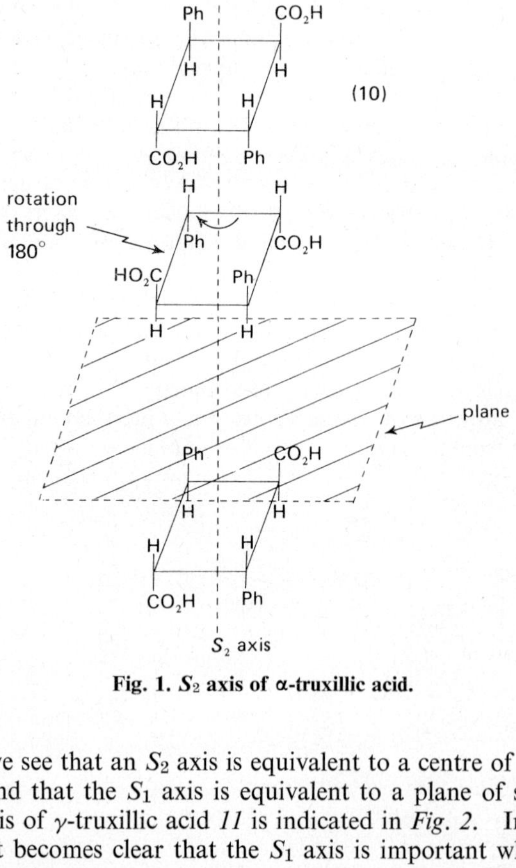

Fig. 1. S_2 axis of α-truxillic acid.

Thus, we see that an S_2 axis is equivalent to a centre of symmetry and we find that the S_1 axis is equivalent to a plane of symmetry. The S_1 axis of γ-truxillic acid *11* is indicated in *Fig. 2*. In terms of chirality it becomes clear that the S_1 axis is important whereas the C_1 axis is not.

Chirality then occurs as a result of dissymmetry or asymmetry and, apart from their reactions with chiral reagents such as enzymes, the enantiomeric molecules differ only in the way in which they interact with plane polarised light and are said to be optically active. One

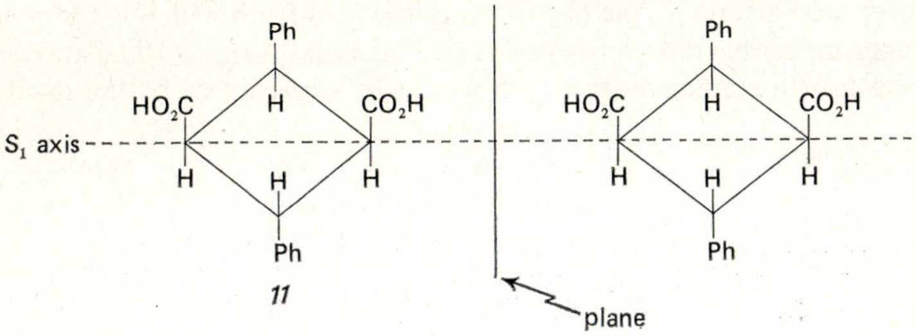

11

Fig. 2. S_1 axis of α-truxillic acid.

enantiomer (liquid or solution) may rotate the plane of polarised light to the left (*laevo* or (−) or *l*) while the other may rotate the plane to the right (*dextro* or (+) or *d*). The rotation is expressed as the **specific rotation** which is given by the equation

$$[\alpha]_\lambda^t = \frac{\alpha_\lambda^t}{l \times \rho \text{ (or } c)}$$

l — sample thickness in decimetres
ρ (or c) — density of the pure liquid (or g ml^{-1} if a solution)
α — observed rotation
t — temperature
λ — wavelength of the light used.

Some solvent dependence is observed and therefore the solvent used should be stated. An equimolar mixture of the *dextro-* and *laevo-* forms shows no optical activity and is called a racemic modification. The optical activity of lactic acid and tartaric acid has already been alluded to on p 1. Lactic acid contains a carbon atom which carries four different groups and the molecule is asymmetric and accordingly exists in the enantiomeric forms *1* and *2*. The carbon atom giving rise to this asymmetry and chirality is referred to as an asymmetric or chiral carbon atom. There are a vast number of molecules containing such a chiral carbon atom, many of which occur naturally, and the α-amino acids are of particular importance. Glycine *12*, the first member of the series, is not chiral as it does not contain a chiral carbon atom but alanine *13* and all the other α-amino acids are chiral and thus have *dextro-* and *laevo-*forms. Although the presence, in a molecule, of a single asymmetric or chiral carbon atom necessarily ensures that the molecule is chiral, in those molecules where two or more chiral carbon atoms are present such overall chirality is not ensured. The maximum number of stereoisomers which are possible in a molecule containing n chiral carbon atoms is 2^n. We have already seen that tartaric acid has only three stereoisomers, the (+)-form *5*, the (−)-form which is the non-superimposable mirror image *14*, and the *meso*-form *6*. The absence of a fourth stereoisomer which would be expected to be the mirror

$H_2NCH_2CO_2H$

12

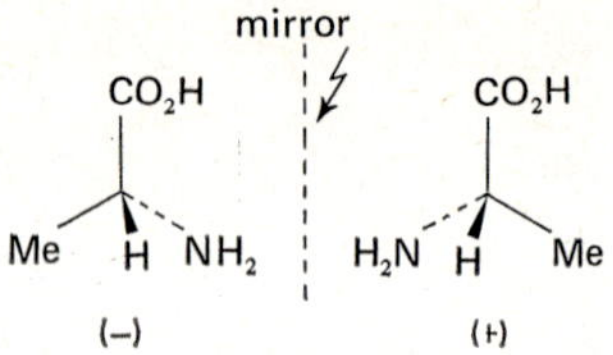

13

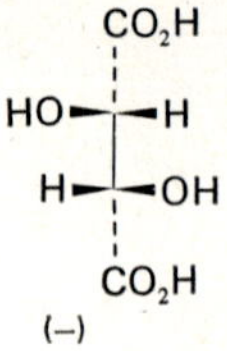

14

image of the *meso*-structure *6* is a result of the presence of a plane of symmetry in this structure. Four stereoisomers are found for the structure *15*, these being the (+) and (−) forms of erythrose *16* and the (+) and (−) forms of threose *17*.

OHCCH(OH)CH(OH)CH_2OH

15

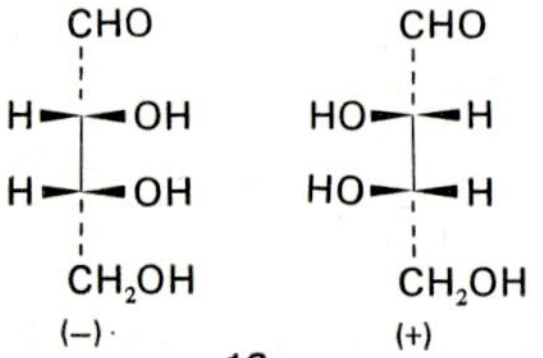

16

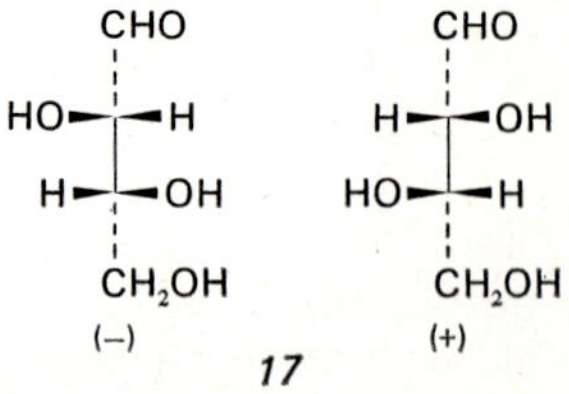

17

These examples show that the presence of more than one chiral carbon atom in a molecule results in stereoisomerism other than enantiomerism. The structure of *meso*-tartaric acid *6* is not the mirror image of either of the optically active forms *5* and *14*. Similarly, there is no mirror-image relationship between the structures of erythrose *16* and threose *17*. Such stereoisomerism is termed **diastereoisomerism** and its use is further exemplified by the observation that α-truxillic acid *10* and γ-truxillic acid *11* are diastereoisomers. Diastereoisomers are thus defined as stereoisomers that do not relate to each other as mirror images and this term includes isomerism owing to the presence of double bonds. For example, *cis*-pent-2-ene *18* and *trans*-pent-2-ene *19* are diastereoisomers.

Traditionally, the term geometrical isomerism has been used to describe this type of stereoisomerism and, although this is still in wide use, it can be seen that it is really superfluous. It is important to note that, unlike enantiomers which differ only in their optical activity and their reactions with chiral reagents, diastereoisomers differ in their physical and chemical properties, usually allowing them to be fairly readily separated by fractional crystallisation or chromatography. It is these differences in physical and chemical properties of diastereoisomers which are in fact responsible for the

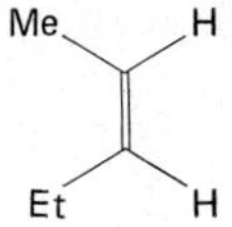

18

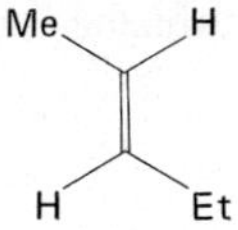

19

differing reactivity of enantiomers with chiral reagents (see p 4). Interaction of a chiral reagent with one enantiomer will produce a transition state which is diastereomeric with the transition state resulting from interaction of the same chiral reagent with the mirror image and thus one would expect one transition state to be favoured.

Thus far the tetrahedral carbon bonds have been represented by wedges and broken lines to indicate their spatial arrangement. It is convenient to use the Fischer projection (*Fig. 3*) which by convention

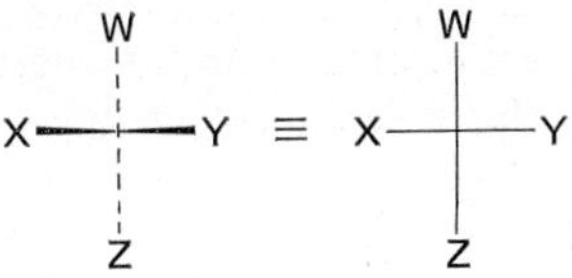

Fig. 3. Fischer projection.

requires that the atoms or groups attached to the horizontal bonds (X,Y) are above the plane of the paper and those attached to the vertical bonds (W,Z) are below the plane of the paper. Thus, (+)-alanine *13* is represented by the projection *20*. How is it known that projection *20* represents (+)-alanine and not the alternative (−)-form? Similarly, although it is known that the relative configurations of the chiral carbons of the (+)- and (−)-forms of tartaric acid are as indicated in structures *5* and *14*, how is it known that structure *5* corresponds to the (+)-form? The determination of absolute configuration was made possible by the X-ray crystal structure analysis of the sodium rubidium double salt of (+)-tartaric acid carried out by Bijvoet in 1951. Although X-ray crystal structure analysis will not normally distinguish between

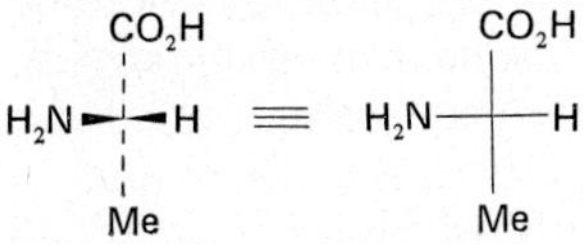

(+) *13* *20*

enantiomers, the incorporation of a heavy atom (in this case rubidium), that will absorb the X-rays, results in anomalous dispersion which allows the determination of the absolute configuration. As a result of Bijvoet's work the absolute configuration of (+)-glyceraldehyde *21* was also established since its configuration relative to that of

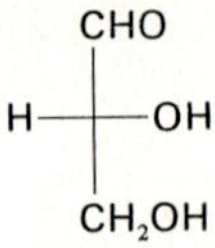

21

tartaric acid was known. Fortunately, this configuration was the same as that arbitrarily chosen by Fischer some 50 years earlier. Absolute configurations are now commonly determined by X-ray crystallographic methods or by chemical correlation with those compounds for which the absolute configuration has been established. Other methods are also used but these are outside the scope of this text.

Although chirality, enantiomers, and diastereoisomers have been discussed in terms of chiral carbon it should be stressed that these terms are in no way restricted to molecules possessing chiral carbon atoms since chirality is a property of molecules as a whole and relates to the absence of certain symmetry elements (σ, S, i). Accordingly, any molecule which, as a result of its bonding geometry or steric deformation, lacks these symmetry elements will be chiral and will have a non-superimposable mirror image.

For example, a substituted allene such as 1,3-diphenylpropadiene *22* will have (+)- and (−)-forms because the planes containing the terminal carbon atoms are orthogonal.

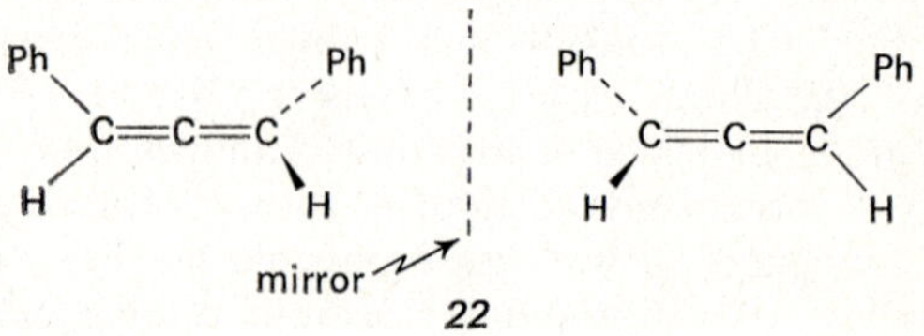

22

In general, allenes of the structure *23* are chiral as they are dissymmetric and possess a chiral axis passing through C=C=C. When $R^1 = R^2$ or $R^3 = R^4$ the molecules become achiral as a plane of symmetry is present. Similar axial chirality is observed for appropriately substituted spirans (*eg 24*) and alkylidene-cycloalkanes (*eg 25*).

A number of substituted biphenyls may also have a chiral axis when the steric (van der Waals repulsive) interactions (*see Chapter 2*)

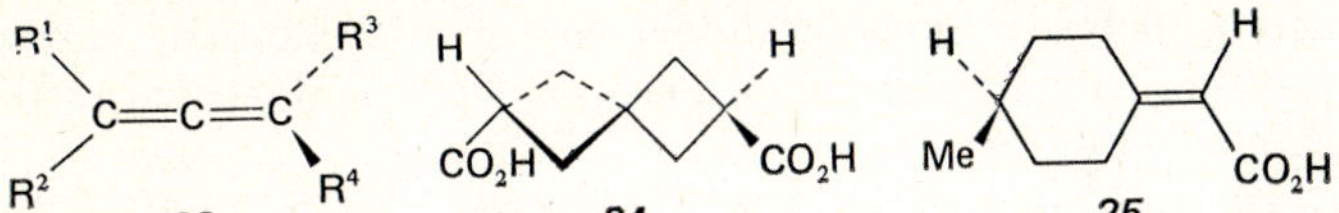

23 24 25

between *ortho*-substituents are sufficient to prevent rotation of the aromatic rings relative to each other. In this situation the preferred conformations (*see Chapter 2*) are those in which the aromatic rings are orthogonal and provided the substitution is of the type in structure *26*, *eg* 6,6′-dinitrobiphenyl-2,2′-dicarboxylic acid *27*, then the minimum energy conformations are enantiomeric. Racemisation of either enantiomer occurs by rotation through 180° about the bond between the aromatic rings and the ease of racemisation depends on the steric compression of the *ortho*-substituents in the conformation in which the aromatic rings are co-planar. In this co-planar transition state some relief of the steric compression will be achieved by molecular deformation, and in particular, the *ortho*-substituents will bend away from each other. Although such deformation requires an input of energy there will be a net gain and, neglecting dipole–dipole interactions, it becomes apparent that the energy barrier to rotation is comprised largely of van der Waals net* repulsion energy and

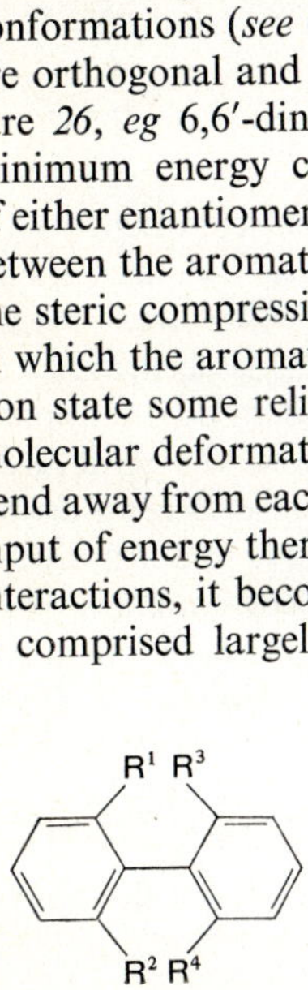

26

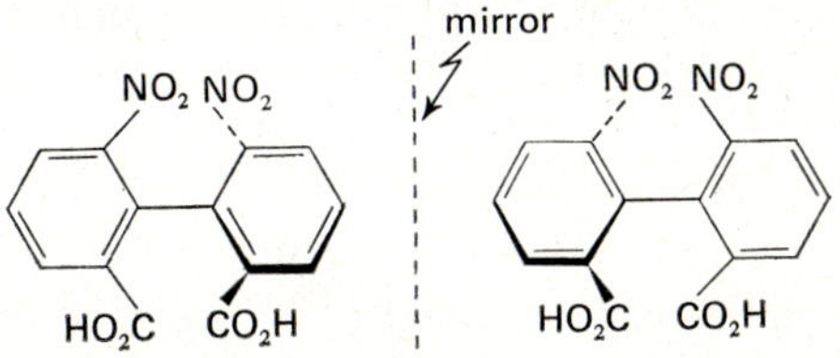

27

molecular deformation energy. Enantiomers of this type are called **atropisomers** and they may be resolved at room temperature when the energy barrier to rotation is of the order of 80 kJ mol^{-1} or more. Clearly when $R^1 = R^2$ or $R^3 = R^4$, structure *26* becomes achiral irrespective of the energy barrier to rotation. In this situation a further substituent at the *meta*-position (but not *para*-) in the aromatic ring carrying the two identical *ortho*-substituents would maintain the dissymmetry of the orthogonal conformation (*eg 28*).

It is interesting to observe here that atropisomers may be described both as configurational isomers, since they are enantiomeric, and conformational isomers, since they are interconverted by rotation around a carbon–carbon single bond. This demonstrates that the

* Allowing for the van der Waals attraction energy (*see Chapter 2*).

distinction between configurational and conformational isomers (see p 1) is not always clear cut. This point will be discussed further in relation to conformation changes (*Chapter 2*).

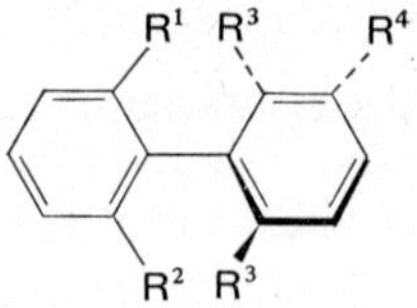

28

From a consideration of allenes it is apparent that cumulenes with an even number of double bonds may have a chiral axis and those with an odd number of double bonds may have *cis*- and *trans*-forms (diastereoisomers). Thus, the general structure *29* is dissymmetric while the structure *30* is not. Similarly, *para*-terphenyls with appropriate substituents may exhibit atropisomerism and the structures *31a* and *31b* are diastereomeric. Although the former has an enantiomer, the latter does not and is a *meso*-compound as it has a centre of symmetry located at the centre of the central aromatic ring (•). Other examples of atropisomerism are known and other chiral structures which do not contain a chiral carbon are also known.

In this latter group are the *ansa*-compounds which result from a special type of restricted rotation and are exemplified by the com-

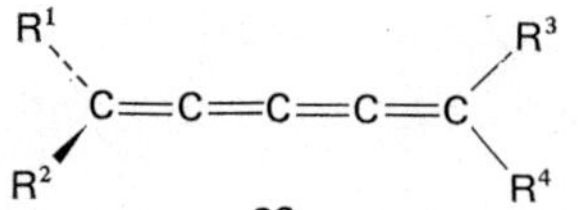

29

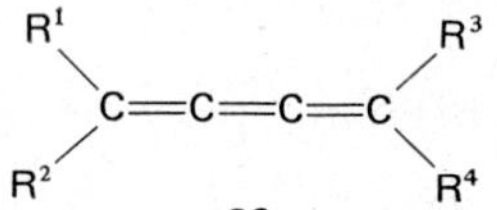

30

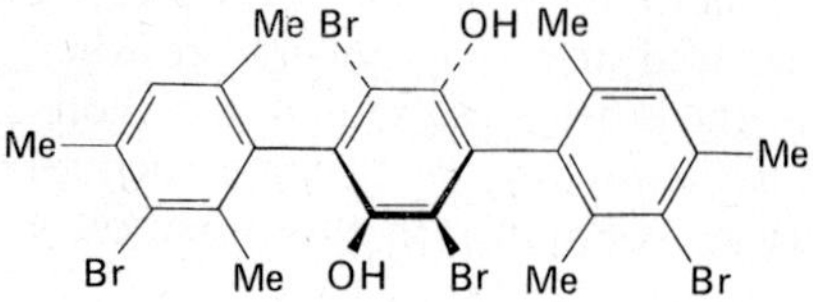

31a

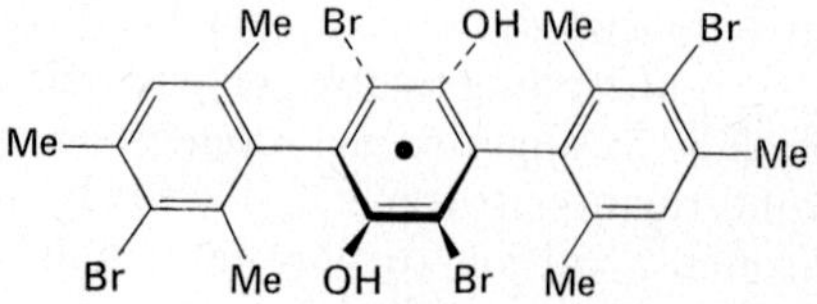

31b

pound *32*. Related to these are the paracyclophanes (*eg 33*) and both of these groups have a chiral plane rather than a chiral centre or axis.

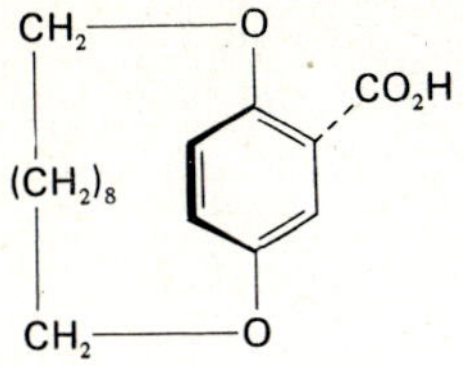

32

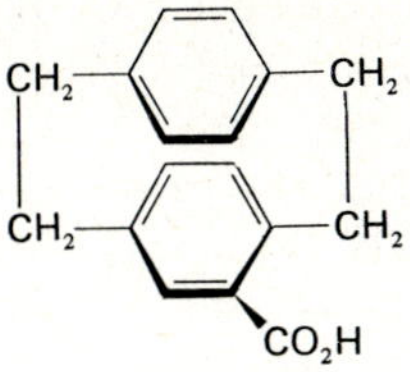

33

The helicenes, *eg* hexahelicene *34*, are a particularly interesting group of chiral molecules. Any helix is inherently dissymmetric and the helical structure of these compounds and therefore their chirality results from steric overcrowding. For example, the planar configuration of hexahelicene *34* is not possible owing to severe steric interactions between the terminal rings. Accordingly these rings move away from each other and overlap giving rise to two helical structures which are enantiomeric. A notable feature of the helicenes is their unusually high specific rotation.

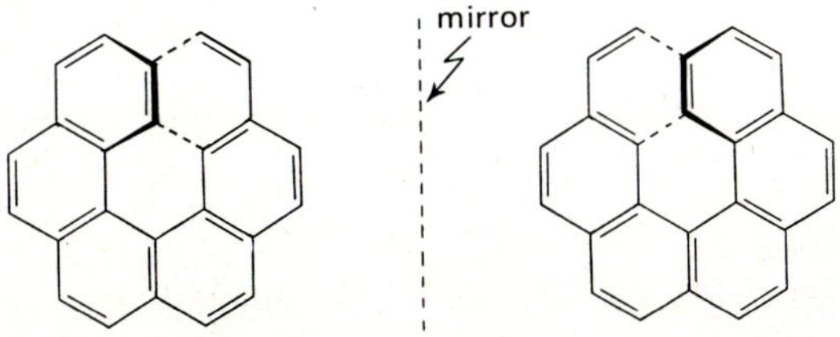

34

Only the occurrence of chirality, enantiomers, and diastereoisomers in carbon compounds has been discussed. It should be stressed that, although outside the scope of this monograph, the stereochemistry of metal coordination compounds has been an area of intense study for many years. The stereochemistry of certain nitrogen-containing organic molecules is particularly relevant to the organic chemist. Quaternary ammonium salts are tetrahedral and compounds such as *35* and the related amine oxides *36* are thus resolvable into enantiomers. Doubly bonded nitrogen compounds such as oximes *37* and imines *38* exhibit diastereoisomerism and have *cis*- and *trans*-forms. As a general rule in such compounds the terms *syn* and *anti* replace *cis* and *trans* respectively.

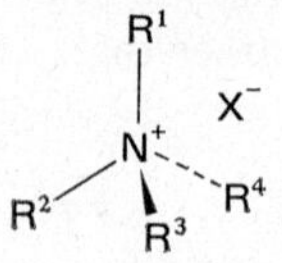

35

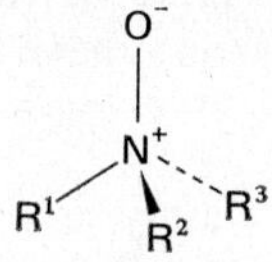

36

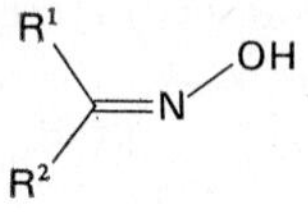

37

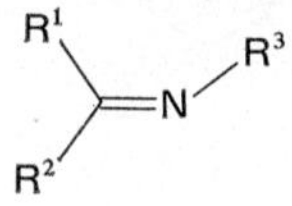

38

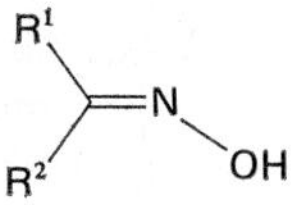

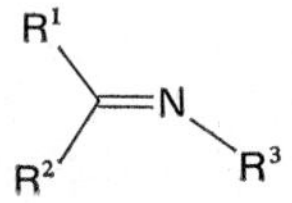

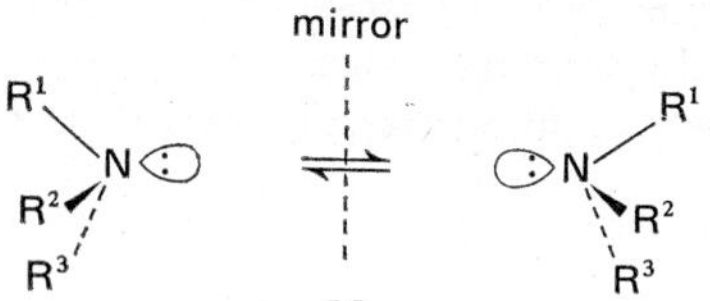

39

Although tertiary amines are pyramidal and the structure *39* is chiral, enantiomers cannot normally be resolved owing to a rapid interconversion between the enantiomers by inversion of the nitrogen. In special circumstances this inversion is slowed down sufficiently to allow the isolation of stereoisomers. For example, the N-chloro-aziridines *40a* and *40b* are separable although they interconvert relatively easily.

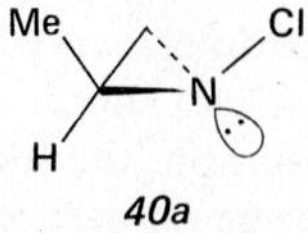

40a

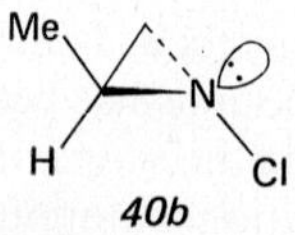

40b

As expected, by analogy with nitrogen-containing compounds, asymmetrically substituted phosphonium salts and phosphine oxides are isolable in enantiomeric forms. Unlike amines, however, phosphines are configurationally stable and thus methylpropyl-phenylphosphine *41*, for example, may be resolved into its enantiomeric forms. Sulphoxides and sulphonium salts may show similar chirality.

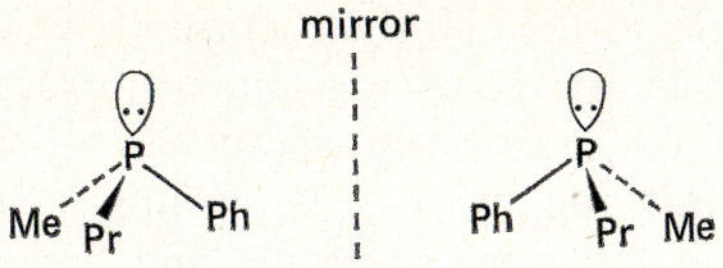

41

Other elements such as boron, silicon, germanium, tin and arsenic are known to give rise to optical activity.

In concluding this section, it is instructive to briefly consider the concept of two dimensional chirality.* Prelog, who was awarded the Nobel Chemistry Prize jointly with Cornforth in 1975 for contributions to the understanding of stereochemistry, has emphasised the importance of this concept.

It has been indicated that *cis*- and *trans*-isomers of alkenes may be described as diastereoisomers, and it should be recognised that they are two dimensional diastereoisomers. In general, compounds such as the *cis*- and *trans*-alkenes *42* and *43* are two dimensional diastereo-

42 43

isomers each having a two dimensional non-superimposable mirror image which is produced by reflection, within the plane of paper, through a straight line which represents a two dimensional mirror. The difference between the two dimensional enantiomers is that they present the two different faces of the planar molecule to the observer. *Cis*-but-2-ene *3* is a two dimensional *meso*-compound whereas *trans*-but-2-ene *4* is two dimensionally chiral. This concept of two dimensional chirality is useful in understanding prochirality and related phenomena which are discussed in the following section.

Prochirality, enantiotopicity, diastereotopicity and pseudoasymmetry

Prochirality and enantiotopicity

Molecules of the type *44*, in which R^1, R^2, and X are achiral, are necessarily achiral and possess a plane of symmetry that passes through the groups R^1 and R^2 and the carbon atom to which these are attached. Replacement of one of the groups X in structure *44*

* In a two dimensional world the simplest chiral figure is then an irregular triangle. The two enantiomers present different faces to the observer.

by a group Y leads to the chiral structure *45* which is enantiomeric with the structure *46* arrived at by a similar replacement of the alternative group X by Y. A carbon atom such as is present in compound *44* is thus **prochiral** and the two X groups have a mirror image relationship within the same molecule and are called **enantiotopic**

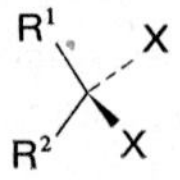

44

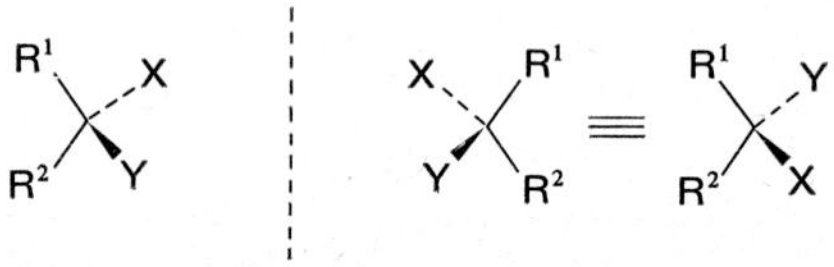

45 mirror *46*

groups. The most significant difference between enantiotopic groups is their reactivity in chiral environments. Enzymes, the chemical catalysts of living systems, being composed of α-amino acids which are chiral, are themselves chiral and will distinguish between enantiotopic groups. It is not difficult to see why this should be so and there is no need to invoke specific requirements for the enzyme-substrate combination such as three point contact (Ogston). Partial bonding of a chiral reagent W (say *dextro*rotatory) with the group X would give rise to two possible diastereomeric transition states *47* and *48* owing to the induced chirality at the

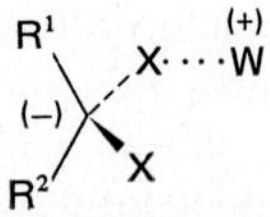

47

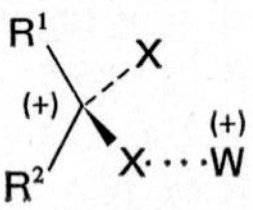

48

original prochiral carbon (arbitrarily assigned (−) and (+)). Since diastereoisomers differ in physical and chemical properties it would be expected that one of the transition states *47* or *48* would be energetically preferred. Accordingly, a chiral reagent would be expected to react selectively with one of the groups X. Obviously, reaction of an achiral reagent with each of the groups X in turn would give enantiomeric transition states which would be energetically indistinguishable and thus the groups are entirely equivalent in their reactions with such achiral reagents.

Consider the oxidation of the two enantiomeric 1-deuterioethanols *49* and *50* which are derived by stereospecific replacement of the

49 *50*

enantiotopic hydrogen atoms in ethanol with deuterium. Oxidation with achiral reagents would, in each case, give an equivalent mixture of acetaldehyde (ethanal) and 1-deuterioacetaldehyde.* However, oxidation of compound *49* with the enzyme alcohol dehydrogenase and the co-enzyme NAD$^+$ gives exclusively 1-deuterioacetaldehyde whereas the enantiomeric species *50*, under the same conditions, gives exclusively acetaldehyde demonstrating the ability of the enzyme to distinguish between the enantiotopic groups. It might be expected that the reverse reaction, namely the enzymic reduction of acetaldehyde to ethanol, would be similarly stereospecific. This is indeed the case and inspection of acetaldehyde shows that it is two dimensionally chiral (see *51* and *52*) and that the two faces of the carbonyl group are enantiotopic. Thus addition of a nucleophile (X^-) to the two faces gives two enantiomeric structures *53* and *54*. Accordingly, reduction of acetaldehyde with NADD/alcohol dehydro-

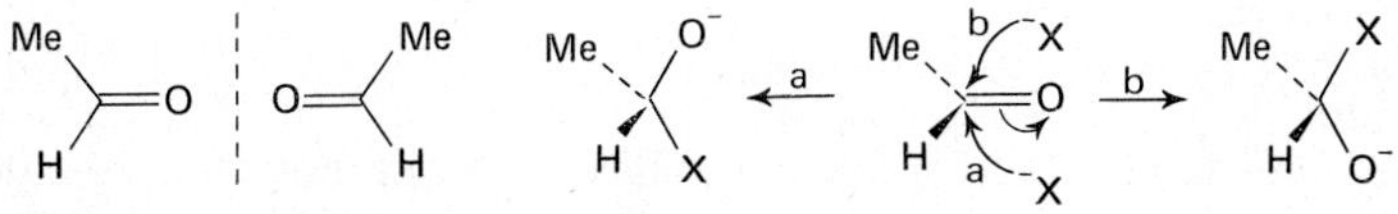

51 *52* *53* *54*

genase gives compound *50* whereas reduction of 1-deuterioacetaldehyde with NADH/alcohol dehydrogenase gives compound *49*.

It is apparent, then, that prochirality is not restricted to tetrahedral carbon structures and returning to the structures *42* and *43*, which are two dimensionally chiral, it can be seen that they have enantiotopic faces. For example, peracid oxidation of compound *42* would give rise to two enantiomeric epoxides *55* and *56*. In the case where $R^1 = R^2$ then structures *55* and *56* become indistinguishable as they are derived from a two dimensionally achiral compound, the faces of which are entirely equivalent. The prochirality of appropriately

* The operation of any kinetic deuterium isotope effect would ensure that an excess of deuterioacetaldehyde would be present.

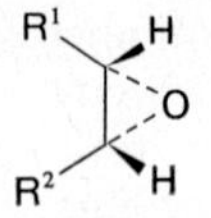

55

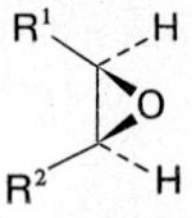

56

substituted alkenes is further demonstrated by the fumarase-catalysed hydration of fumaric acid (*trans*-butenedioic acid) *57* to the malic acid (hydroxybutanedioic acid) *58* rather than its enantiomer. Not surprisingly, maleic acid (*cis*-butenedioic acid) *59*, which is two dimensionally achiral, exhibits no prochirality and the two faces of the double bond are equivalent.

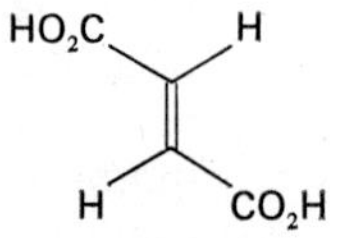

57

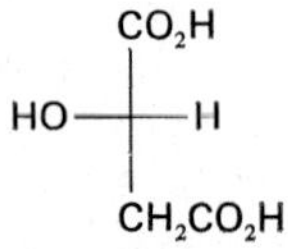

58

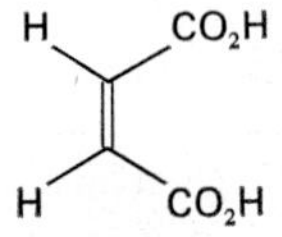

59

It is apparent that there is a link between two dimensional chirality and prochirality in the normal, three dimensional world. Fumaric acid *57* has a C_2 axis which is perpendicular to the plane of the molecule and which bisects the double bond. Rotation through 180° around this axis provides an identical structure, but it does not interchange *57* with its two dimensional enantiomer *60*: it does not interchange the two faces of the molecule (*see Fig. 4*). Maleic acid *59* has a C_2 axis which is within the plane of the molecule and which bisects the double bond (*see Fig. 4*). Rotation through 180° around this axis does interchange the two faces of maleic acid and thus the molecule is not prochiral since the two faces are indistinguishable.

In general, then, a planar molecule which in a two dimensional world would be chiral will exhibit prochirality. In these circumstances it lacks a C_2 axis within the plane of the molecule *viz* the

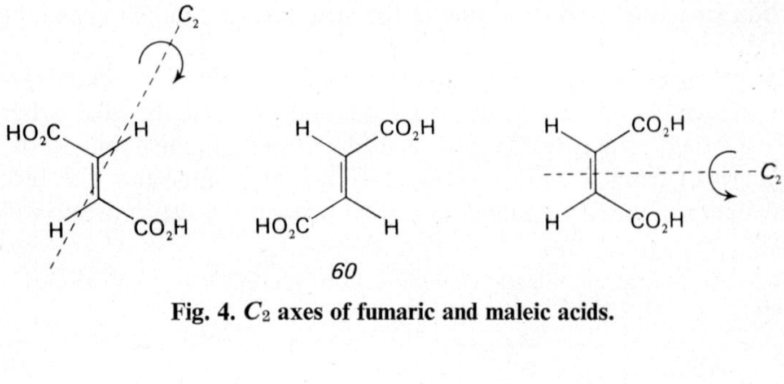

60

Fig. 4. C_2 axes of fumaric and maleic acids.

plane of symmetry. In tetrahedral carbon compounds the same principle applies.

In structure *44* there is no C_2 axis within the plane of symmetry of the molecule and the two faces of the plane are enantiotopic as are the groups X on either side of the plane. In a structure such as *61*, however, there are two planes of symmetry, each of which having a common C_2 axis. Accordingly, the carbon is not prochiral and there

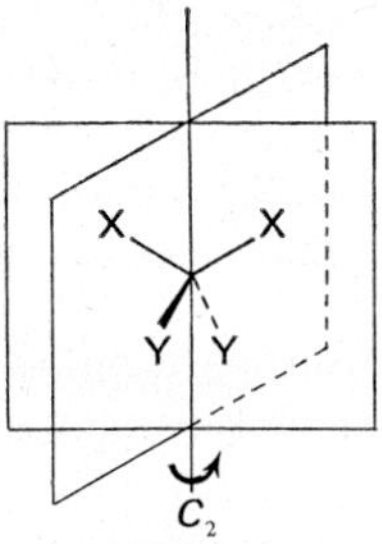

61

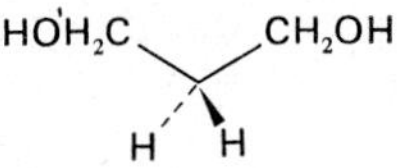

62

are no enantiotopic groups since rotation through 180° interchanges the faces of the planes of symmetry. Propane-1,3-diol *62* is an example of such a molecule.

Diastereotopicity

Consider now the situation in structure *44* when either R^1 or R^2 is chiral. The groups X in the new structure that arises (*eg 63*) are now **diastereotopic** since replacement of one of these with a group Y gives rise to a diastereoisomer *64* of the structure *65* which is produced by

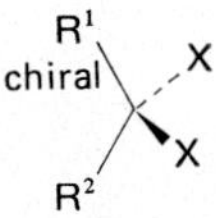

63

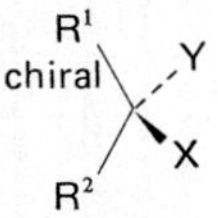

64

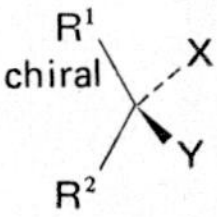

65

replacing the alternative group X with Y. Unlike enantiotopic groups, diastereotopic groups are distinguishable by achiral as well as chiral reagents. This is relatively easily understood by reference to the structures *66* and *67* (*cf 47* and *48*), which represent diastereomeric transition states arising from interaction of the reagent W with the different X groups. The induced chirality in structure *66* is, of course, opposite to that in structure *67* and the chirality of R^1 (rather than W) ensures that the transition states are diastereomeric and have

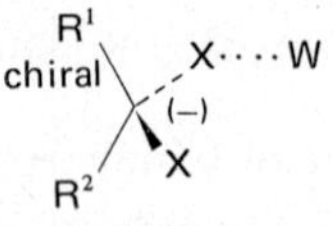

66

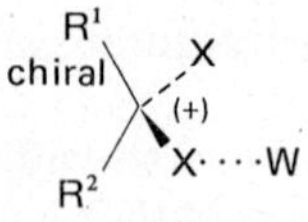

67

MeCF(Cl)CH_2Cl

68

different energies. Diastereotopic groups are quite often distinguishable in the nmr spectrum.

An example which illustrates the general principle of diastereotopicity and demonstrates the non-equivalence of diastereotopic groups in the nmr spectrum is 1,2-dichloro-2-fluoropropane *68* in which the protons of the methylene group are diastereotopic and give rise to two overlapping quartets (δ3.6–4.2) in the ^{1}H nmr spectrum (*Fig. 5*) owing to their being spin–spin coupled to each other and the fluorine atom.

It has been demonstrated that the addition of a hydride ion (and other nucleophiles) to unsymmetrical carbonyl compounds using chiral reagents provided a single enantiomer stereospecifically owing to the faces of the carbonyl group being enantiotopic. If the carbonyl group is attached to a chiral group as in structure *69*, then the faces are diastereotopic and addition of a hydride ion to the two faces, from an achiral reagent, provides diastereoisomers *70* and *71*

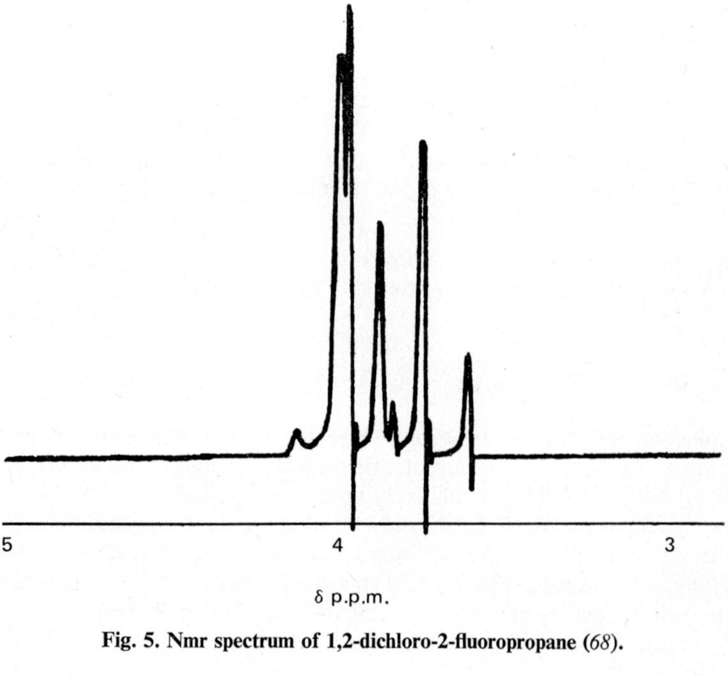

Fig. 5. Nmr spectrum of 1,2-dichloro-2-fluoropropane (*68*).

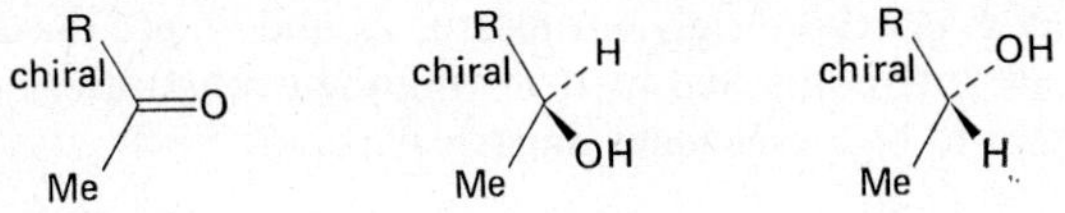

69 70 71

through diastereomeric transition states. Accordingly, it would be expected that one of the products would be produced in excess. This is an example of asymmetric induction and the Cram rule (*see Chapter 5*) allows the prediction of the predominant isomer in many cases.

Pseudoasymmetry

Certain molecules contain carbon atoms which are attached to two identical chiral groups, which may or may not have the same absolute configuration, and two further different achiral groups. Structure *72* represents a general example in which the absolute configuration of the chiral R group is not specified. In the simplest cases where R has one chiral structural feature there will be four stereoisomers and perhaps the best known examples of these are the trihydroxyglutaric acids *73*, *74*, *75* and *76*.

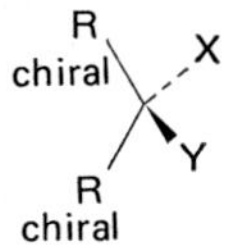

72

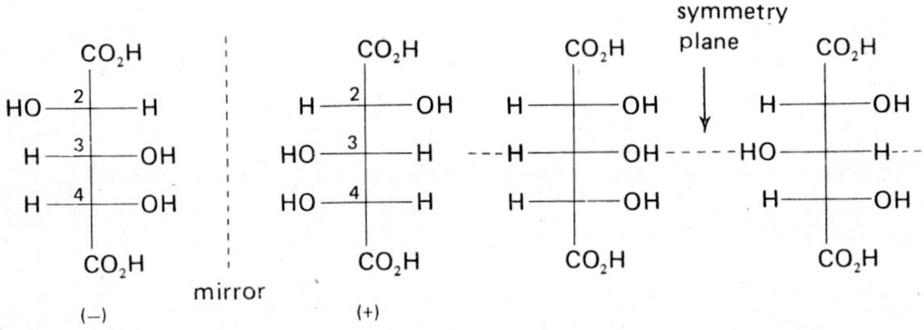

73 74 75 76

Structures *73* and *74* are enantiomeric, despite C-3 being achiral, and the absolute configurations of C-2 and C-4 are the same in the former and, of course, enantiomeric with those of C-2 and C-4 in the latter. However, structures *75* and *76* are achiral since they possess a plane of symmetry owing to the configurations of C-2 and C-4 being enantiomeric. Accordingly, structures *75* and *76* are diastereomeric and they are interconverted by inverting the configuration of the C-3 which is said to be a **pseudoasymmetric** carbon.

Configurational nomenclature

It is obviously necessary, as a matter of convenience, to be able to describe the configuration of a chiral molecule by an unambiguous symbol rather than have to draw three-dimensional perspective figures or the simpler two-dimensional Fischer projections. As a result of Rosanoff's modifications of Fischer's original attempts at classification of carbohydrates, the D- and L-nomenclature has evolved. Although this convention has been largely superseded by the *R*- and *S*-nomenclature based on the sequence rules of Cahn, Ingold and Prelog (see p 21), it is still used particularly for carbohydrates and α-aminoacids. Fischer projections are used to assign the symbols D or L. Accordingly, (+)-glyceraldehyde *21* is arbitrarily defined as D because the OH group attached to C-2 is on the right hand side of the Fischer projection in which the CHO group appears at the top. Its enantiomer is defined as L because the OH group is on the left hand side. In the carbohydrates, in general, the position of the OH group

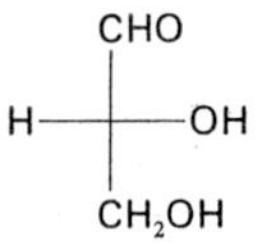

D-(+)

21

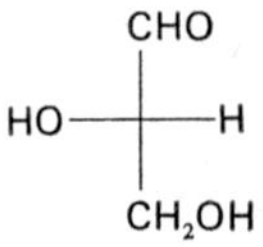

L-(−)

77

attached to the penultimate carbon atom in the chain determines the assignment of D or L. Thus (+)-glucose *78* and (−)-fructose *79* have the D-configuration and (+)-ribose *80* has the L-configuration.

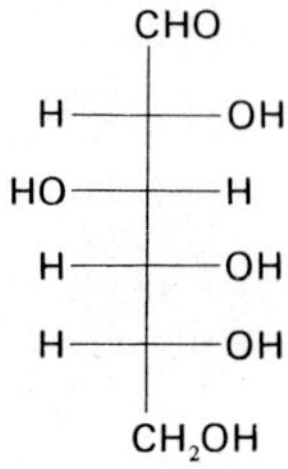

78

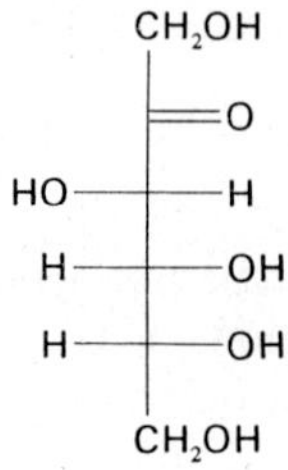

79

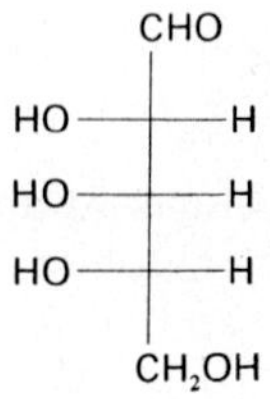

80

Similarly, (−)-erythrose *16* has the D configuration and (+)-threose *17* has the L configuration.* It should be stressed that the symbols D and L do *not* relate to the sign of rotation of an optically active molecule which is designated (+) (or *d*) and (−) (or *l*). In the α-aminoacids the L-compounds are those in which the NH_2 group is on the left hand side of the Fischer projection in which the carboxyl group appears at the top. Conversely, the D-compounds are those in which the NH_2 group is on the right hand side. Thus (+)-alanine *20* and (−)-serine *81* are L-α-aminoacids. Although the D and L

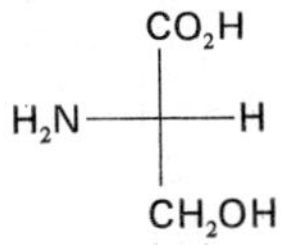

81

nomenclature appears satisfactory for carbohydrates and α-aminoacids, it suffers from serious deficiencies when applied in general to molecules containing more than one chiral carbon. For example, (−)-tartaric acid *14* may be classified as belonging to the D-series if it is considered to be derived from a carbohydrate. Alternatively, if it is considered to be derived from an α-aminoacid it would be classified as belonging to the L-series. An additional problem may arise in the D and L system of nomenclature when assigning the configuration of a molecule which differs markedly chemically from any of the standards. For example (+)-glyceraldehyde *21* could be classified as D or L since it may be correlated with L-(−)-serine *81*, or its enantiomer (D), because the molecules differ in two chemical groups. In general, then, assignment of the configurational symbols D or L will not automatically allow a complete three dimensional model to be constructed.

The *R* and *S* system, devised by Cahn, Ingold and Prelog, does offer an unambiguous method of defining absolute configuration. Correlation with an arbitrary standard configuration is not involved. Using a set of rules — the sequence rules — the order of priority of groups surrounding a chiral carbon is determined. If, when viewing the molecule from opposite the group of lowest (fourth) priority the remaining groups, in decreasing order of priority, describe a clockwise motion then the configuration is described by the symbol *R* (*rectus*). Alternatively, when the remaining groups, in decreasing order of priority, describe an anticlockwise motion then the con-

* The terms erythro and threo are derived from these tetroses and are used to describe diastereomeric structures in molecules containing two chiral carbon atoms of the type R^1 Cab Cac R^2. The erythro form is that in which the groups a are on the same side in the Fischer projection and the threo form is that in which the groups are on opposite sides.

figuration is described by the symbol *S* (*sinister*). Accordingly, in structure *82* in which the sequence rules define the order of priority of the groups as say a > b > c > d then the configuration is *R*. Conversely, in structure *83* the configuration is *S*.

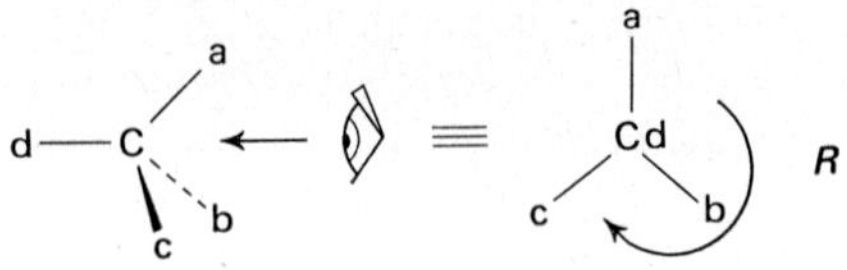

82

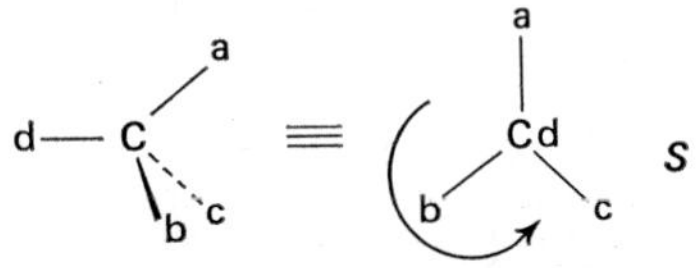

83

The sequence rules which determine the order of priority of the groups may be summarised as follows:

- Atoms of higher atomic number take precedence over those of lower atomic number, thus I > Br > Cl > F > O > C > H, and isotopes of higher atomic weight take precedence over those of lower atomic weight, thus $^{3}_{1}H > ^{2}_{1}H > ^{1}_{1}H$. Lone pairs are assigned the lowest possible priority.
- When the atoms attached directly to the chiral carbon are the same, then the priorities are determined by the atomic numbers of the atoms which are directly attached, in turn, to these. The atoms of highest atomic number in each group are compared followed by the next highest and so on until a difference is observed. Thus $CH_2Cl > CH_2OH > CH_3$ because Cl > OH > H and $(CH_3)_3C > (CH_3)_2CH > CH_3CH_2$ because C > H. If no difference is observed for this second set of atoms, then the third set or the fourth and so on is considered in the same manner until such a difference is arrived at.
- In groups containing a multiple (double or triple) bond then for the purposes of determining priority the multiple bond is duplicated or triplicated at both ends. For example, –CHO becomes

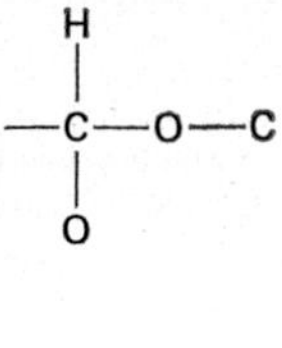

and takes precedence over $-CH_2OH$, $-C\equiv N$ becomes

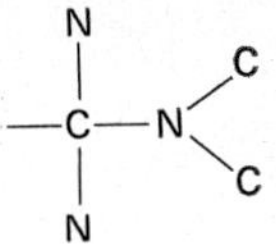

and takes precedence over $-CH_2NMe_2$, and

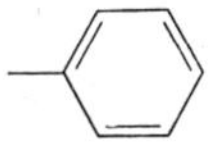

becomes

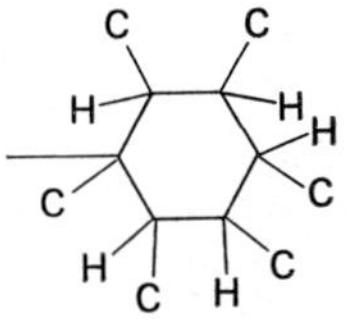

and takes precedence over cyclohexane.

- When the difference between substituents is in configuration then $R > S$ and *seqcis* > *seqtrans* (see p 25).

In (+)-glyceraldehyde *21* the order of priority of the groups is $OH > CHO > CH_2OH > H$ and the configuration is *R* (*Fig. 6*).

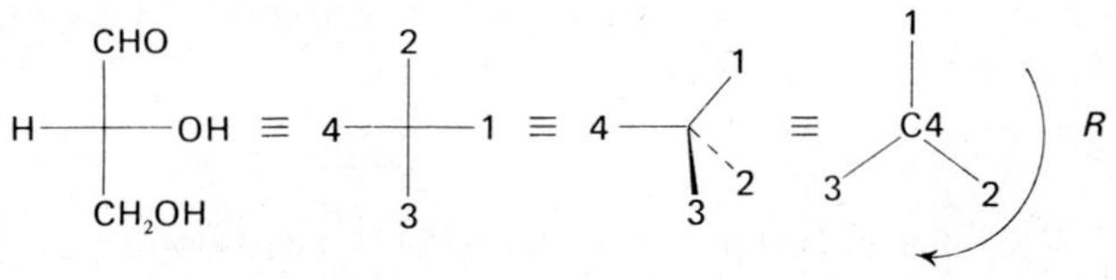

21

Fig. 6. Configurational symbol for (+)-glyceraldehyde.

Similarly in (−)-serine *81*, the order of priority of the groups is $NH_2 > CO_2H > CH_2OH > H$ and the configuration is *S* (*Fig. 7*). In the tartaric acids, and in other molecules containing more than one chiral carbon, the *R* and *S* assignments are made for each chiral carbon. Thus (−)-tartaric acid *14* can be shown to have the 2*S*, 3*S*-configuration (*Fig. 8*).

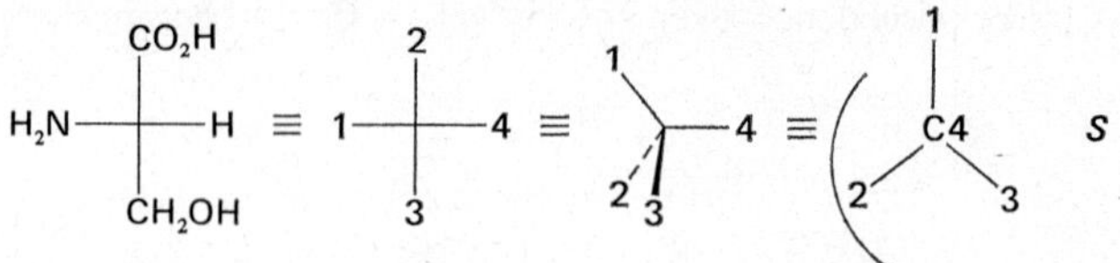

81

Fig. 7. Configurational symbol for (−)-serine.

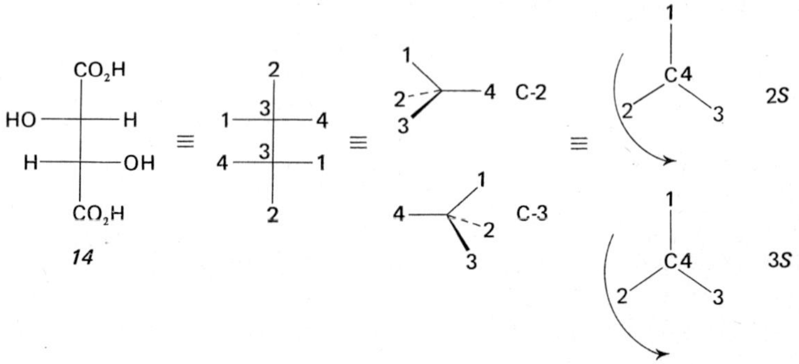

14

Fig. 8. Configurational symbols for (−)-tartaric acid.

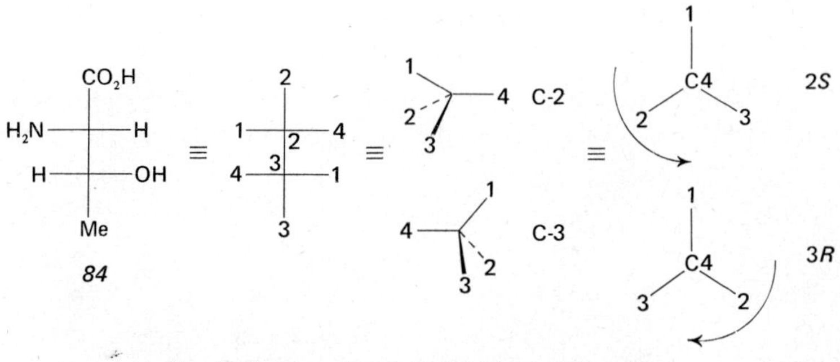

84

Fig. 9. Configurational symbols for (−)-threonine.

Similarly (−) threonine *84* has the 2*S*, 3*R*-configuration (*Fig. 9*).

It is instructive to consider the application of the *R* and *S* system to the *meso*-trihydroxyglutaric acids *75* and *76* in which the configurations of the pseudoasymmetric C-3 differ. The configuration of C-2 in *75* and *76* is *R* while that of C-4 is *S*, therefore since $R > S$ we might expect to assign the symbol *R*. In fact, since structure *75* is achiral and C-3 is a pseudoasymmetric carbon the symbol *r* is employed (*Fig. 10*). Similar arguments lead to the conclusion that C-3 in structure *76* is *s* (*Fig. 10*).

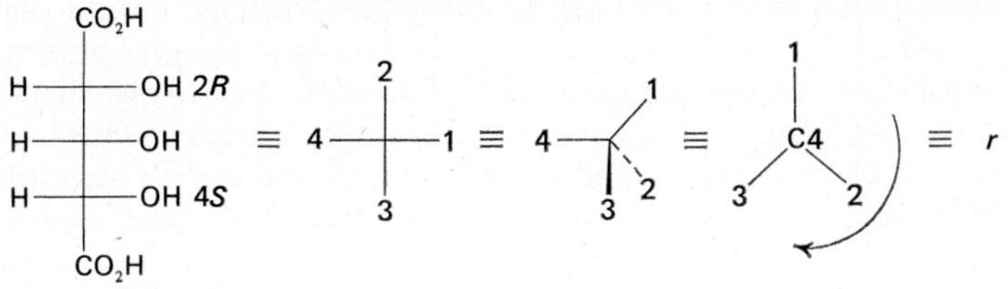

75

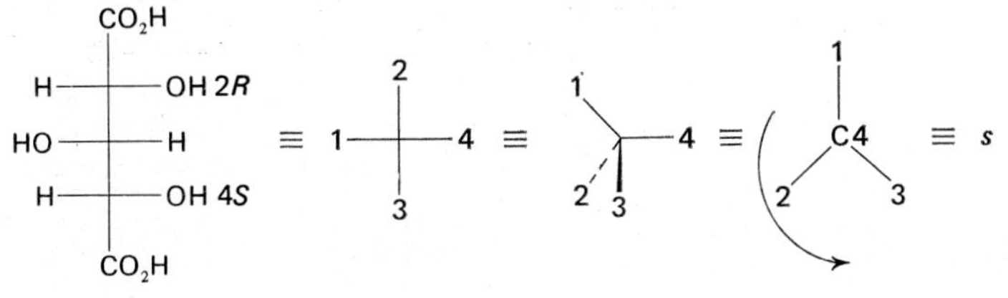

76

Fig. 10. Configurational symbols for *meso*-trihydroxyglutaric acids.

The terms *cis* and *trans* have been used to describe the diastereomeric forms of simple alkenes (see p 1). However, the use of these terms is not always unambiguous and recently, to avoid ambiguities which may arise, the sequence rules have been employed to assign the configuration of alkenes. When the two groups of higher priority from each end of the double bond are on the same side of the double bond then the configuration is defined as *seqcis* or *Z* (from the German *zusammen*, together) and when the two groups of higher priority are on opposite sides of the double bond then the configuration is defined as *seqtrans* or *E* (from the German *entgegen*, across). Thus, 1,2-dichloro-1-bromoethylene exists in the *seqcis* or *Z* form *85* and the *seqtrans* or *E* form *86*.

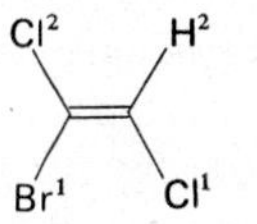

85

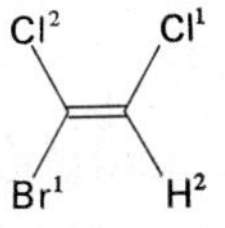

86

The sequence rules may also be used to designate the absolute configurations of molecules containing a chiral axis or a chiral plane and the principles involved in their application to the former are illustrated by consideration of 6,6′-dinitrobiphenyl-2,2′-dicarboxylic acid *27*. The molecule is viewed along the chiral axis and the four substituents giving rise to the chirality are projected onto a plane at right angles to the chiral axis. The resultant projection is now regarded as representing a chiral carbon with the proviso that the near groups always take precedence over the far groups and the bonds between the near groups and the imaginary chiral carbon are always above the plane of the projection. Thus, the enantiomer shown in *Fig. 11* has the *S* configuration, and it is clear that it is unimportant

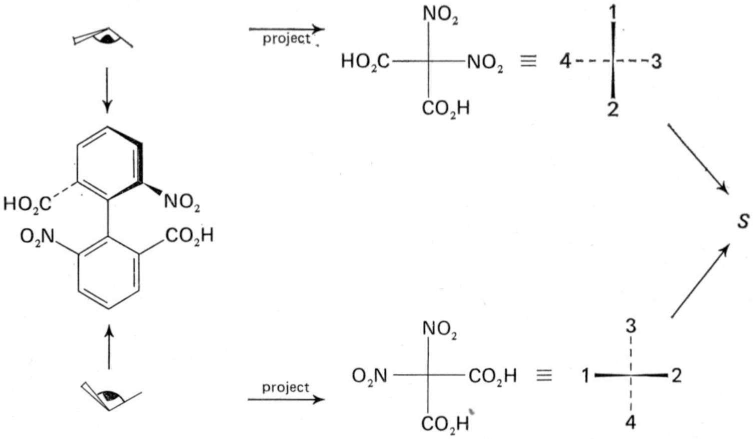

Fig. 11. Configurational symbol for (−)-6,6′-dinitrobiphenyl-2,2′-dicarboxylic acid.

from which end the molecule is viewed. Mislow was able to show (1957) that this enantiomer was the (−)-form of 6,6′-dinitrobiphenyl-2,2′-dicarboxylic acid.

The sequence rules may be applied to distinguish between enantiotopic (and between diastereotopic) groups. The general principles may be illustrated with ethanol *87*. The enantiotopic hydrogen atoms are designated *pro-S* and *pro-R* by the appropriate subscripts. Replacement of the *pro-S* hydrogen atom by a group W of higher precedence than itself but of lower precedence than the other two different groups gives the structure *88* with the *S* configuration, while replacement of the *pro-R* hydrogen atom by the group W gives the structure *89* which has the *R* configuration. More generally, in the

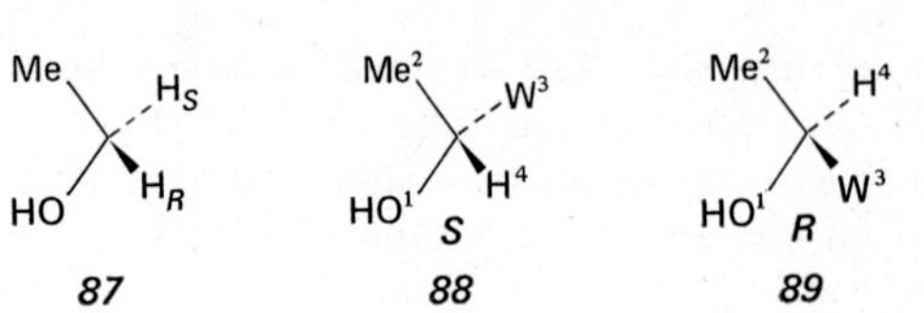

system CR^1R^2XX *44* one of the groups X is elevated in priority over the other so that the priorities relative to the groups R^1 and R^2 remain unchanged. When the resultant chiral carbon has the *R* configuration then the group X, which has been so elevated, is designated *pro-R*. Conversely, if the derived chiral carbon has the *S* configuration then the elevated group X is designated *pro-S*. Application of these principles to the diastereotopic hydrogen atoms of glyceraldehyde lead to the designations shown in *90*.

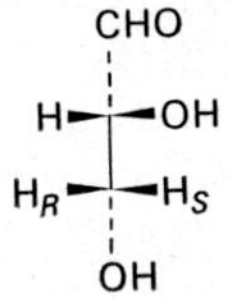

90

2. Conformation

The conformations of a molecule are the different spatial arrangements of the atoms or groups which result from rotation around a single bond and, in the most general sense, these different conformations may be regarded as stereoisomers (*see Chapter 1*). Each different conformation will have a different energy and the lower energy conformations will be populated in preference to those of higher energy. Where a molecule has a number of conformations which correspond to energy minima these are called **conformational isomers** or **conformers**. It is not implied by these terms that such conformations are necessarily isolable, though in certain cases this can occur. For example, the conformational isomers of some *ortho*-substituted biphenyls, which are at the same time configurational isomers, may be isolated. In these cases the energy barrier to rotation around the bond joining the aromatic rings is caused principally by strong van der Waals repulsive interactions between the *ortho*-substituents. The following discussion will illustrate other examples of conformational isomerism and demonstrate the importance of torsional (Pitzer) strain and, in certain cases, bond angle (Baeyer) strain, in addition to van der Waals strain.

Energy considerations in simple acyclic molecules

The conformation of ethane changes as one methyl group rotates relative to the other. The two extremes, in terms of energy, are the staggered conformation *91*, which has the minimum energy, and the eclipsed conformation *92*, which has the maximum energy. Both may be conveniently represented as Newman or saw-horse projections. A Newman projection is a projection onto the page of the view along

Newman projection

Saw-horse projection

91 *92*

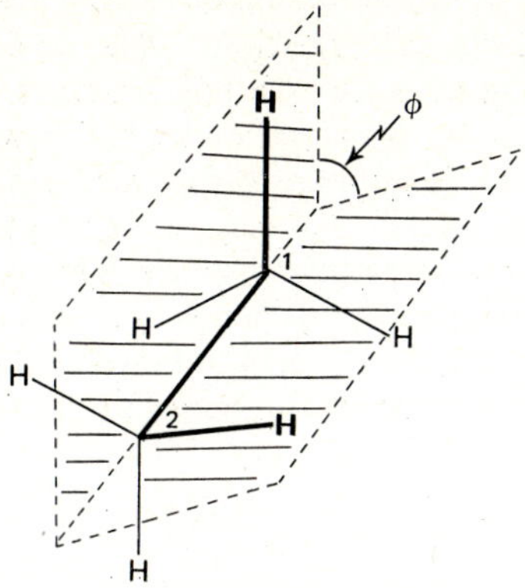

Fig. 12. Dihedral angles in ethane.

a carbon–carbon bond; the intersecting lines represent the nearer carbon and its three remaining bonds, and the circle and its radial lines, the farther carbon and its three remaining bonds. In the saw-horse projection the molecule is viewed from an angle.

The differences between conformations *91* and *92* is readily explained in terms of the dihedral or torsional angle (ϕ) which is the angle between planes containing H–C-1–C-2 and H–C-2–C-1 (*Fig. 12*). Since there are three equivalent hydrogen atoms on C-1 and C-2 it is necessary to specify one hydrogen atom on C-1 and C-2. The dihedral angle for the staggered conformation is 60° while that for the eclipsed conformation is 0° and *Fig. 13* shows the variation of potential energy* with dihedral angle for ethane. There is an energy barrier to rotation of *ca* 12.5 kJ mol^{-1} and in one complete revolution there are three equivalent eclipsed and three equivalent staggered conformations. This energy barrier was determined by Pitzer and Kemp (1936) in their studies on the heat capacity and entropy of ethane. However, its source is still a matter for discussion. The separation of a hydrogen atom on C-1 from its nearest neighbour on

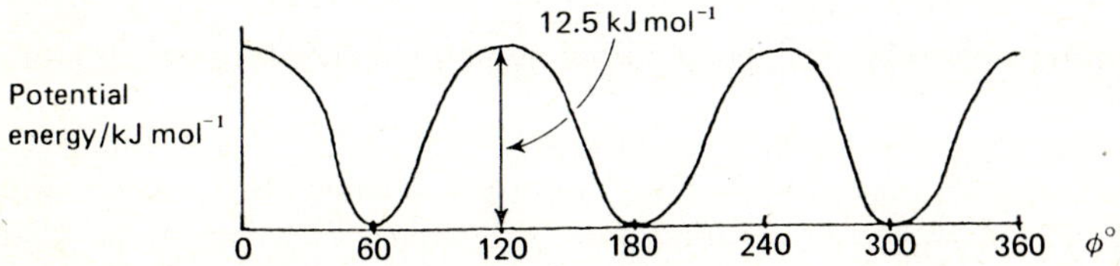

Fig. 13. Ethane: potential energy change with torsional angle.

* The change in potential energy may be equated with the standard enthalpy change, $\Delta H^{\ominus}$

C-2 is *ca* 2.3 Å (1Å = 10^{-10} m) in the eclipsed conformation and *ca* 2.5 Å in the staggered conformation. The van der Waals energy of interaction (E_V) between such non-bonded pairs of atoms may be estimated using the following equation by Hill in which the first term relates to the relatively long-range attractive forces (London) and the second term relates to the relatively short-range repulsive (steric) interaction; r is the interatomic distance between the atoms concerned, r^* is the sum of their van der Waals radii and ϵ is an energy parameter which determines the depth of the van der Waals minimum of energy (*Fig. 14*) and depends on the sizes of the atoms involved.

$$E_V = -2.25\epsilon\left(\frac{r^*}{r}\right)^6 + 8.28\,(10^5)\,\epsilon\,\exp\left(\frac{-r}{0.0736r^*}\right)$$

Calculations using a range of van der Waals radii (*ca* 1.2–1.6 Å) and ϵ values for hydrogen indicate that the staggered conformation of ethane is more stable than the eclipsed conformation but only a minor part of the energy barrier to rotation can be ascribed to the van der Waals interactions.† The major part of the energy barrier which

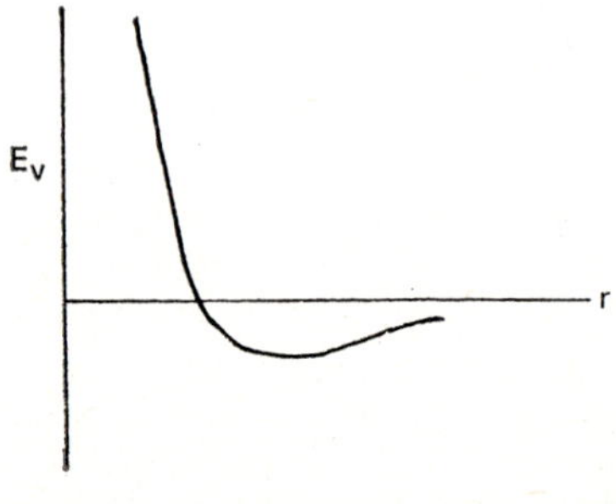

Fig. 14. Variation of van der Waals energy with interatomic distance.

is called the **torsional strain** is not readily understood although a number of theories have been advanced. One possibility is that the rotation barrier may be ascribed essentially to the π-molecular orbitals of ethane‡ which are more bonding in the staggered conformation than in the eclipsed conformation. Although no clear picture emerges as to the cause of the major part of the energy barrier to rotation, its existence is not in doubt and ethane largely adopts the preferred staggered conformation. It should be stressed that the

† It is not known to what extent the discrepancy relates to the inadequacy of this application of the Hill equation which was originally derived for non-polar spherical non-bonded atoms or simple molecules.

‡ See W. L. Jorgensen and L. Salem, *The organic chemist's book of orbitals*, New York: Academic Press, 1973.

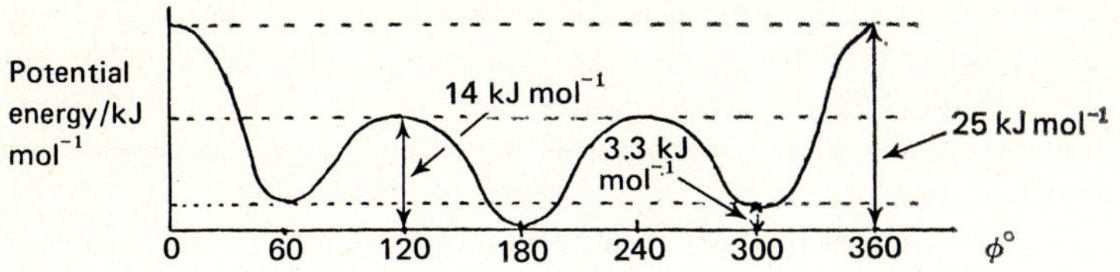

Fig. 15. Butane: potential energy change with torsional angle.

energy barrier to rotation in ethane is insufficient to prevent rotation of one methyl relative to the other as this would require an energy barrier of around 80 kJ mol^{-1} (*see Chapter 1*). The change in torsional strain from zero to maximum as the vicinal bonds move from the staggered to the eclipsed arrangement in ethane is a general phenomenon which has profound conformational effects.*

In the case of butane the change of energy with torsional angle which is defined as the angle between the Me–C-1–C-2 and Me–C-2–C-1 planes, is rather more complex than in the case of ethane (*Fig. 15*). There are two energy barriers to rotation and two minimum energy conformations (conformers) these being the *anti 93* and the skew or *gauche 94*. The maximum energy conformations which

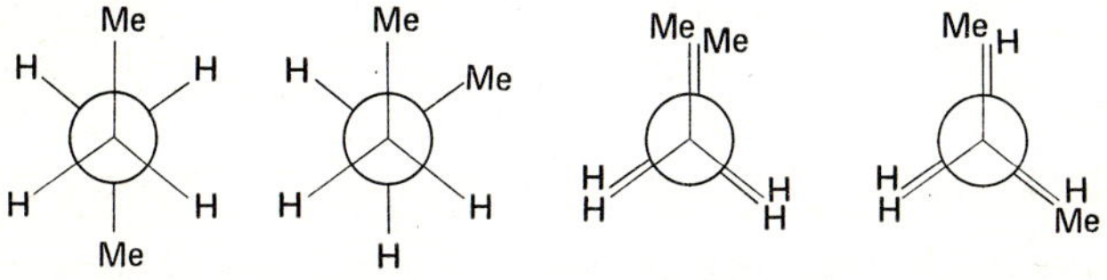

93 *94* *95* *96*

define the energy barriers to rotation are the fully eclipsed *95* and the eclipsed *96*.† The higher barrier to rotation arises from van der Waals net repulsive interactions, principally between the eclipsed methyl groups, and torsional strain owing to the eclipsing of three pairs of vicinal bonds. This torsional strain is present in the lower energy barrier to rotation which is approximately equal to the barrier in propane. The van der Waals strain‡ is considerably reduced and is very little different from that in ethane. There is no torsional or van der Waals strain in the *anti* conformer and the small energy difference between this and the *gauche* conformer (3.3 kJ mol^{-1}) is

* These concepts are, of course, not restricted to organic molecules. The dihedral angle of the minimum energy conformation of hydrogen peroxide is 94°. Although hydrogen peroxide is isoelectronic with ethane, its preferred conformation is different owing to the presence of lone pairs of electrons which repel each other more than bonding electron pairs.

† Other terms used to describe the conformations *93*, *94*, *95*, and *96* are respectively *anti*-periplanar, *syn*-clinal, *syn*-periplanar, and *anti*-clinal.

‡ van der Waals strain indicates a net van der Waals repulsive interaction.

ascribed to net van der Waals repulsion between the two *gauche* methyl groups. The precise torsional angle for the *gauche* conformer is probably slightly greater than 60° so as to afford some relief from the steric interactions without greatly increasing the torsional strain.

The precise geometry of the conformers of butane, allowing for various bond angle and bond length deformations, and the significance of the resultant van der Waals and torsional strain is a matter of current discussion. A number of force field calculations have been reported and it has been suggested (Allinger) that *gauche* H:H interactions, rather than being unimportant, as implied above, could be of major conformational importance. There is, however, some dispute over this interpretation and for the purposes of this text, it will be assumed that the *gauche* H:H interactions are not sterically very significant. Similarly, and as implied above, *gauche* Me:H interactions are also sterically not very significant.

Butane preferentially adopts the *gauche* and *anti* conformers in the approximate ratio 1:2 at room temperature. The relatively high population of the *gauche* conformers relates to there being two such conformers ($\phi = 60°$ and 300°) compared with only one *anti* conformer in one complete revolution. The equilibrium constant K for two conformers may be determined from the general equation

$$\Delta G^{\ominus} = -2.303\,RT \log_{10} K$$

where $-\Delta G^{\ominus}$, the conformational free energy, is equal to the excess of standard free energy of one conformer over that of the minimum energy conformer, and $\Delta G^{\ominus} = \Delta H^{\ominus} - T\Delta S^{\ominus}$, $\Delta H^{\ominus}$ being the standard enthalpy difference and $\Delta S^{\ominus}$ the standard entropy difference between the conformers. The $T\Delta S^{\ominus}$ term may often be neglected, thus $\Delta G^{\ominus} \approx \Delta H^{\ominus}$. However, in the case of butane, owing to the increased probability of the *gauche* conformer *vs* the *anti* conformer, $T\Delta S^{\ominus}$ becomes significant and $-\Delta G^{\ominus}$ is reduced to 1.7 kJ mol^{-1}. Interestingly, these two *gauche* conformers are enantiomeric but not resolvable since the energy barrier to their interconversion is only 14 kJ mol^{-1}.

In general, non-polar molecules will adopt those conformations in which the torsional and van der Waals strains are minimised. The situation becomes rather more complicated for molecules which possess two or more polar groups or atoms. In these molecules, the total energy of any particular conformation may be affected markedly by dipole–dipole interactions or by hydrogen bonding such that the energetically preferred conformations could not be predicted from a consideration of torsional and van der Waals interactions alone.

For example, gaseous 1,2-dichloroethane is more stable in the *anti* conformer *97* than the *gauche* conformer *99* by *ca* 4.6 kJ mol^{-1} owing largely to dipole–dipole repulsion in the latter. It is believed that there is very little steric compression between the chlorine atoms in *99* and the overall van der Waals interactions are probably attractive rather than repulsive. In the condensed phase, dipole–dipole repulsions decrease and, in the pure liquid, conformers *97* and *99* have approximately the same enthalpy. In 2-chloroethanol it

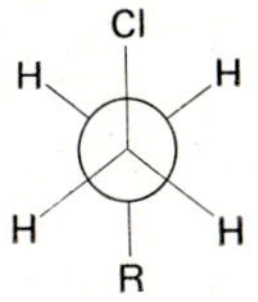

97 R = Cl
98 R = OH

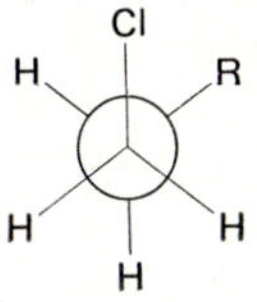

99 R = Cl
100 R = OH

would be anticipated from a consideration of van der Waals and dipole–dipole interactions that the *anti* conformer *98* would be more stable than the *gauche* conformer *100* in the gas phase. In fact, it is estimated that the latter is more stable by *ca* 4 kJ mol^{-1} owing to hydrogen bonding and it is the form in which 2-chloroethanol crystallises.

Conformations of carbocyclic molecules

Cyclopropanes and cyclobutanes

Cyclopropane *101* necessarily has a single planar conformation in which the pairs of vicinal C—H bonds are eclipsed. Formally the C—C—C bond angle would be 60° assuming that the bonds were formed along the internuclear axes. Such a bond angle, which differs markedly from the normal tetrahedral bond angle of sp^3 hybridised carbon would necessarily involve large angle (Baeyer) strain. In real terms, the bonds are not deformed in this way and the angle strain relates to a different orbital arrangement which is of a higher energy than that expected for the formation of simple σ bonds. Qualitatively, the molecular orbitals may be regarded as being formed from imperfectly sp^3 hybridised carbon atomic orbitals such that the orbitals forming the C—C bonds have greater *p* character than sp^3 orbitals while those forming the C—H bonds have correspondingly greater *s* character. Such an arrangement would give rise to bent or 'banana' bonds in which the orbital overlap takes place outside the internuclear axis (*Fig. 16*). This rationalisation accounts for some of the properties of cyclopropane including the H—C—H bond angle of 118° and the C—H bond length of 1.08 Å which are intermediate between the values for methane and ethylene (ethene). There are alternative more precise descriptions of the molecular orbitals of cyclopropane *101*.*

However, irrespective of the precise orbital arrangements in cyclopropane and other structures, the concept of angle strain is a useful one.

In cyclobutane *102* the angle strain is considerably less than in cyclopropane *101* but in a planar conformation there would be considerable torsional strain from the eclipsing of the vicinal pairs of

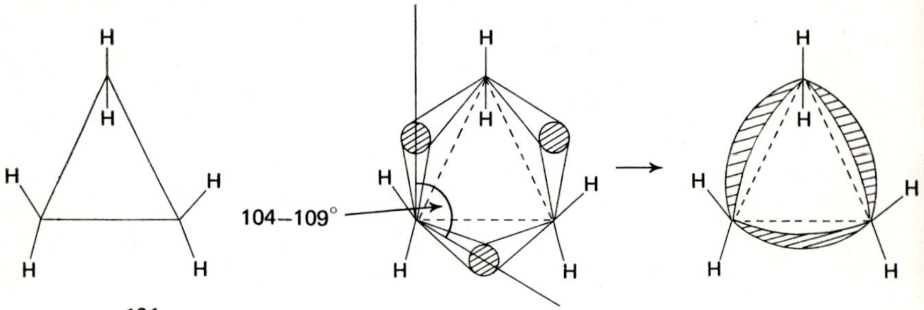

101

Fig. 16. Bent bonding in cyclopropane.

C—C and C—H bonds. The puckered conformation *103* is energetically preferred since it relieves considerably the torsional strain while only marginally increasing angle strain. The degree of puckering in various cyclobutane derivatives depends on the nature of the substituents.

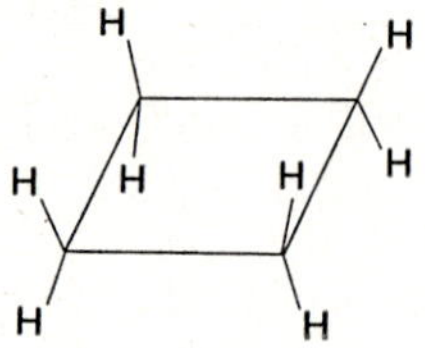

102

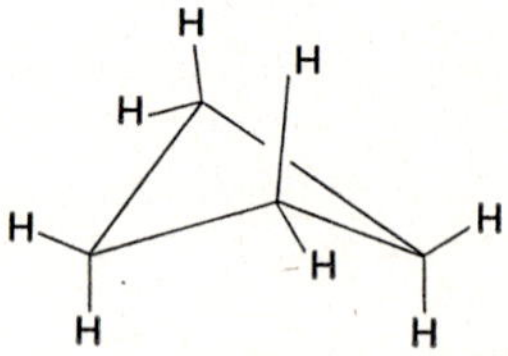

103

Cyclopentanes

The situation in cyclopentane *104* is not dissimilar from cyclobutane. A planar conformation would involve very little angle strain as the 'bond angle' for a regular pentagon (108°) differs only slightly from the normal tetrahedral bond angle. However, such a conformation would suffer considerable torsional strain and, in order to relieve

* W. L. Jorgensen and L. Salem, *The organic chemist's book of orbitals.* New York: Academic Press, 1973.

this, cyclopentane assumes a puckered conformation *105* in which the angle strain is slightly increased, but in which there is a net gain in energy. The puckering of the ring is not fixed and the methylene group which is out of plane rotates around the ring. This so-called 'pseudorotation' does not involve a very great change in energy and no definite energy minima or maxima are apparent.

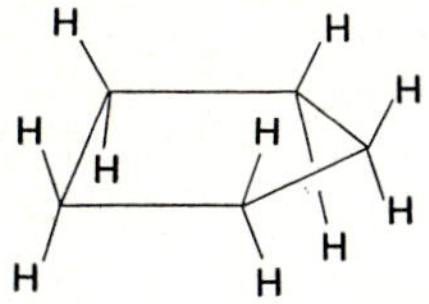
104

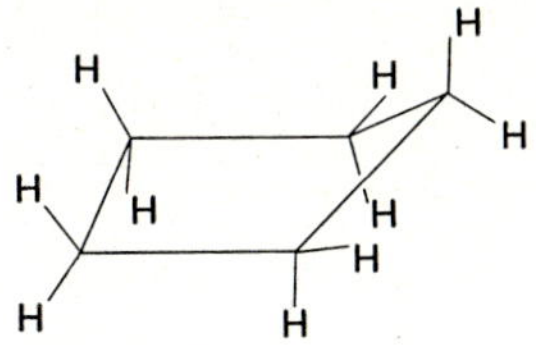
105

In certain heavily substituted cyclopentanes and cyclopentanones better defined conformations are reported but these appear to be exceptional (*see Chapter 4*).

Cyclohexanes

Unlike cyclopentane *104*, cyclohexane *106* has important minimum energy conformations (conformers), these being the chair or rigid conformer *8* and the flexible conformer, neither of which are planar. The latter describes a number of conformations, the two extremes of which are the boat *107* and the skew-boat *108*. The variation of potential energy with conformation is shown in *Fig. 17* and the maximum energy conformation is believed to be a half-chair *109*.

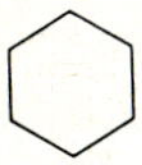
106

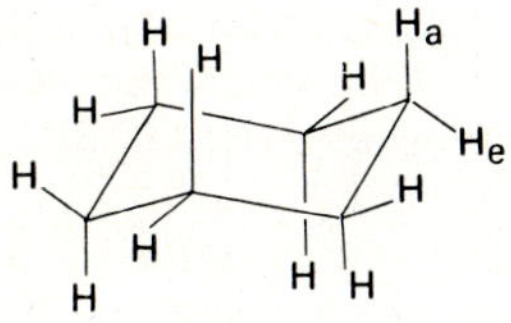
8

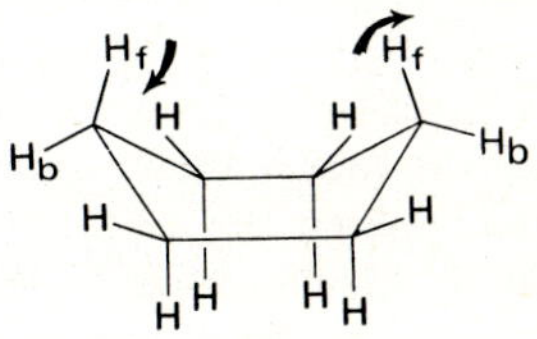
107

e – equatorial, a – axial, f – flagpole, b – bowsprit

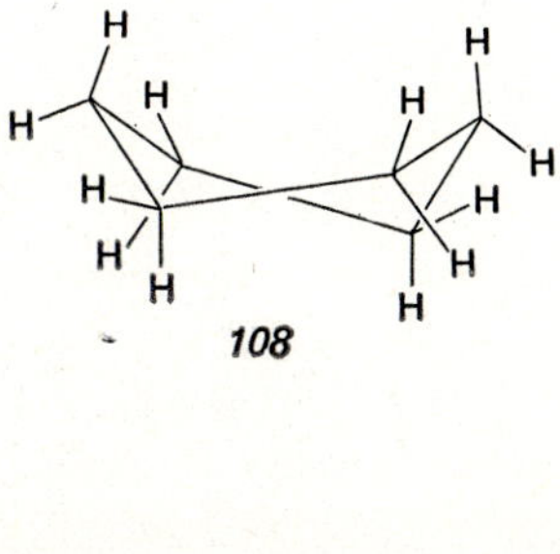
108

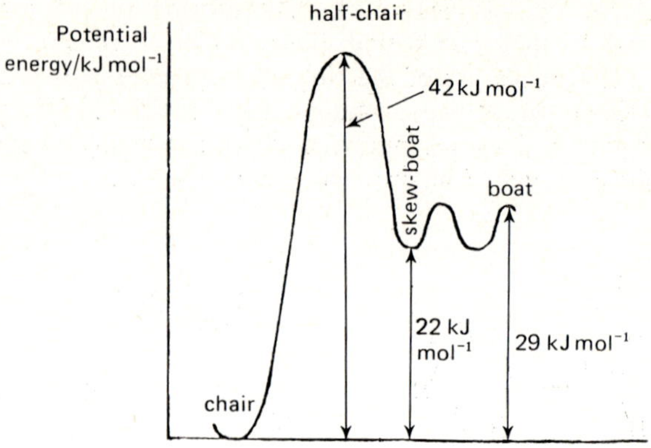

Fig. 17. Variation of potential energy with conformation for cyclohexane.

The interconversion of the chair and flexible conformers involves rotation around the carbon–carbon bonds and some bond angle deformation and the energy barrier of *ca* 42 kJ mol^{-1} is readily surmounted at ambient temperatures.

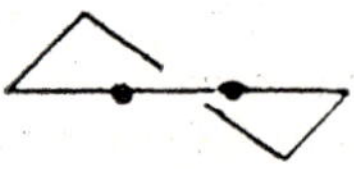

109

The chair and flexible conformers, unlike the planar and half-chair conformations, have no angle strain and the chair has no torsional or van der Waals strains. A Newman projection *110* involving any two adjacent carbon atoms in the chair conformer clearly demonstrates that the vicinal C—H and C—C bonds are fully staggered. The increased energy of the boat conformer arises from van der Waals strain between the so-called flagpole hydrogen atoms on positions 1 and 4, which are *ca* 1.8 Å apart, and from torsional strain owing to the eclipsed pairs of hydrogen atoms on positions 2 and 3 and on 5 and 6 (*see 111*). Some relief of the torsional and van der Waals

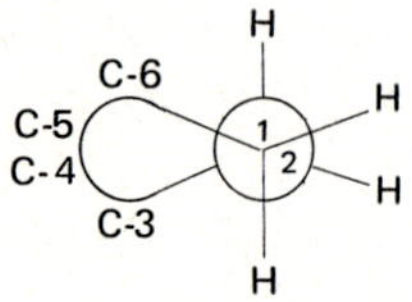

110

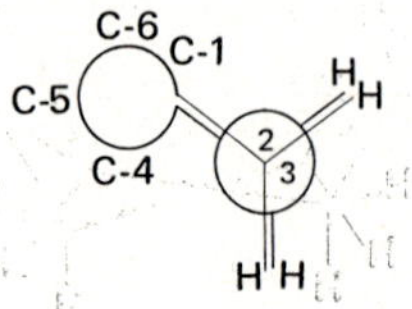

111

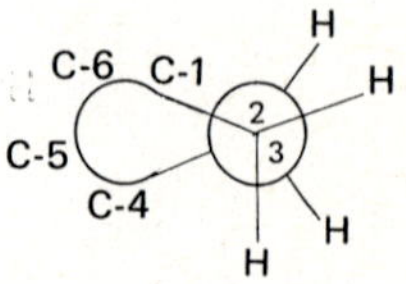

112

strains in the boat conformer is afforded by the flagpole hydrogen atoms moving away from each other as indicated by the arrows in *107* and this leads to the skew-boat (*108*) which has the Newman projection *112*. It is clear that cyclohexane adopts the chair conformer preferentially, and at 25°C it is estimated that about one molecule in a thousand* is in the flexible conformer which is intermediate in the chair–chair interconversion (ring inversion). There are two distinct kinds of C—H bonds in the chair conformer, these being axial and equatorial. As indicated in *8*, the axial bonds are parallel to the C_3 axis while the equatorial bonds make an angle of 109° 28′ with this axis. The boat conformer *107* has similar equatorial and axial bonds and bowsprit bonds in addition to the flagpole bonds which have already been mentioned.

Simple substituted cyclohexanes generally adopt the chair conformer preferentially and monosubstituted cyclohexanes may exist in two conformations in which the substituent is equatorial or axial.

Thus, methylcyclohexane *113* consists essentially of an equilibrium mixture of the conformers *114* and *115* in the ratio *ca* 19:1 respectively. Conformer *114* is favoured energetically because the axial

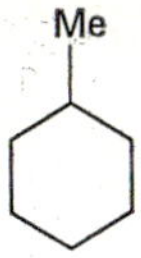

113

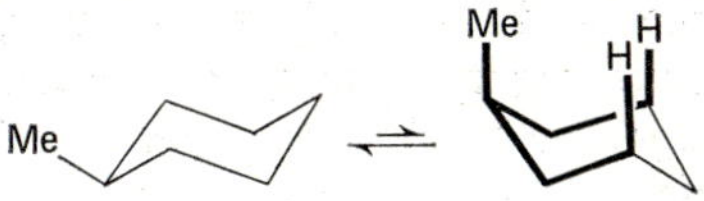

114 115

methyl group of *115* suffers a net van der Waals repulsive interaction with the axial hydrogen atoms at positions 3 and 5. This van der Waals repulsion which may be equated with two *gauche* Me:Me interactions of butane — as indicated by the heavy bonds in *115* — is absent in *114* which has only the insignificant *gauche* Me:H interactions of propane. The enthalpy difference between *114* and *115* would thus be expected to be *ca* 2 × 3.3 kJ mol^{-1} and this figure is in good agreement with experimental values. Owing to the methyl group being a spherical group it may be assumed that the entropy differences between *114* and *115* are negligible and that the conformational free energy† is approximately the same as the enthalpy difference ($-\Delta G^\ominus = 6.6$ kJ mol^{-1}).

In general, in the axial conformer of monosubstituted cyclohexanes, the substituent suffers steric interactions with the axial

* The favourable entropy of the flexible *vs* the chair conformer contributes significantly.

† See J. A. Hirsch in *Topics in stereochemistry*, ed E. L. Eliel and N. L. Allinger. New York: Wiley, 1967, vol 1, p 199 for table of conformation free energies (A values) of substituted cyclohexanes.

hydrogen atoms on the same side of the ring (*syn*-1,3-diaxial interactions) even though the *syn*-axial hydrogen atoms of cyclohexane *106* do not sterically interact.

These unfavourable interactions arise because of the parallel arrangement of the axial bonds and a plan view *116* of a monosubstituted cyclohexane demonstrates that, as the size of the substituent increases and provided the bond length between it and the ring carbon remains constant,* the steric interaction increases. The increase in size of an equatorial substituent is sterically much less critical as it is staggered with regard to the vicinal hydrogen atoms

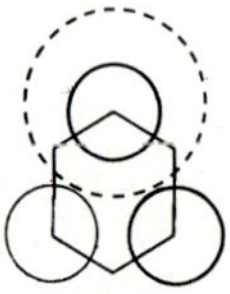

116

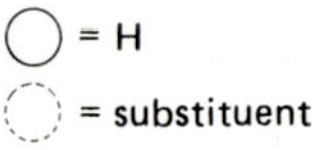

117

and the equatorial bonds diverge *117*. It has been suggested, on the basis of force field calculations (Allinger, *see* p 32), that the preference for a tertiary hydrogen to be axial, which affords relief of important *gauche* H:H interactions, is of overriding importance in determining the conformational preferences of substituted cyclohexanes. This explanation, however, is disputed and for the purposes of this text it will be assumed that *syn*-1,3-diaxial interactions are the major factors involved in determining conformer stabilities. As the size of the substituent increases it is apparent that the equatorial conformer will become more favoured and *t*-butylcyclohexane *118* exists essentially as the conformer *119*.

Even so, it must be stressed that although the axial conformer *120* is undetectable the interconversion between it and *119* still takes place at ambient temperature and *t*-butylcyclohexane *118* and many simple derivatives are conformationally biased molecules rather than conformationally fixed.

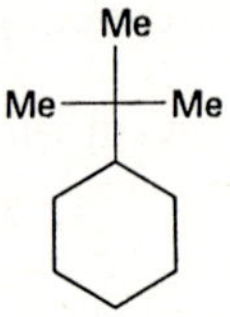

118

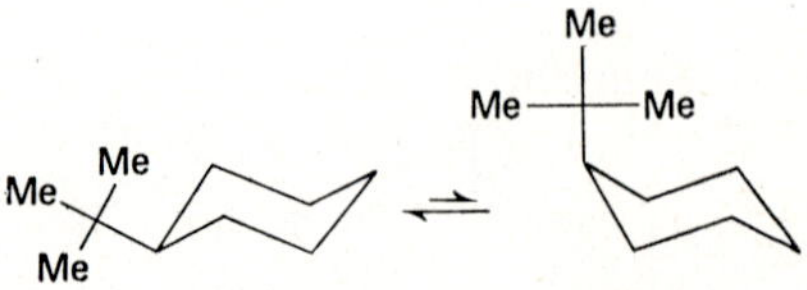

119 120

* If this bond length increases the steric interaction is reduced and *vice versa.*

The conformations of di- and poly-substituted cyclohexanes will, in the absence of polar effects such as dipole–dipole interactions and hydrogen bonding, be determined by steric effects. Thus, for example, *trans*-4-*t*-butylcyclohexanol *121* will have the diequatorial conformer *122* and the *cis*-compound *123* will have the conformer *124* in which the bulky *t*-butyl group is equatorial. The alternative conformations diaxial for *121* and equatorial (OH), axial (*t* Bu) for *122* are much less favoured energetically. The importance of polar

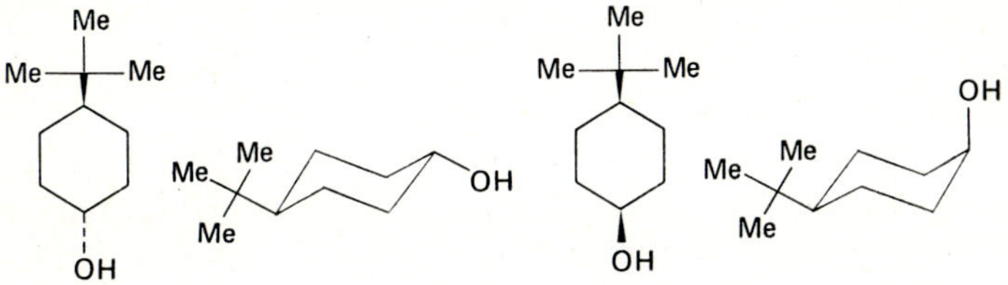

121 *122* *123* *124*

groups in determining conformational preferences has previously been alluded to for acyclic molecules (*see* p 32) and is exemplified here by consideration of *trans*-1,2-dibromocyclohexane *125* which from steric arguments alone would be expected to adopt the diequatorial conformation *126*. In carbon tetrachloride at 30 °C it appears that the conformers *126* and *127* are in the ratio *ca* 1:2 owing to

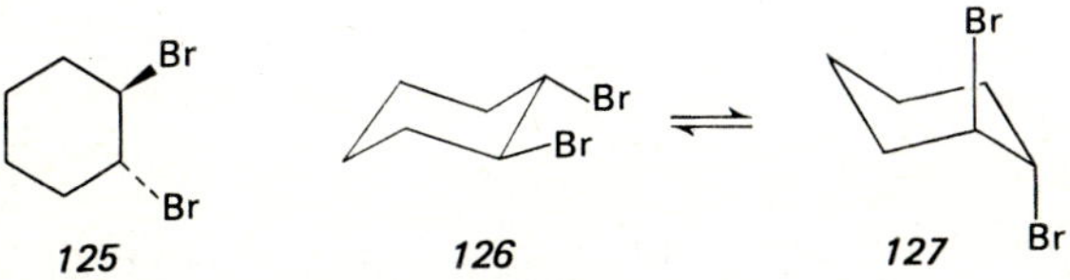

125 *126* *127*

the unfavourable dipole–dipole repulsion in *126*. It is interesting that although bromine has a similar van der Waals radius to a methyl group it is much less sterically hindered in the axial conformation owing to the relatively long ring carbon–bromine bond (see earlier). Additionally, the bromine atom would be expected to be more polarisable than the methyl group and thus the attractive term in the van der Waals equation would be expected to be relatively more important.

It is clear that substituted cyclohexanes will not have precise chair conformations as the molecules will distort to relieve any steric or polar constraints. For example, *syn*-1,3-diaxial interactions lead to a flattening of the chair conformation and although this increases the torsional and angle strain there is a net gain in energy. In certain cases bulky axial substituents can force the molecule to adopt the flexible form. For example, *trans*-1,3-di-*t*-butylcyclohexane *128* exists in the flexible form because the chair conformation *129* would

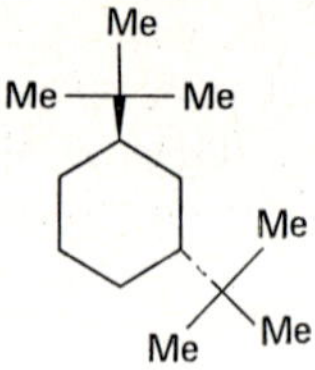

128

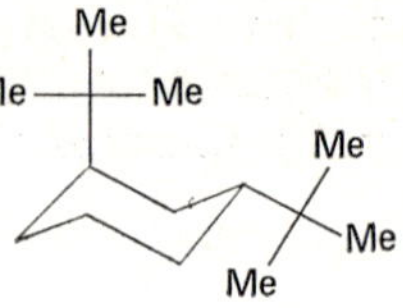

129

necessarily require one of the *t*-butyl groups to be axial. Although it is often satisfactory when discussing the chemistry of cyclohexanes to assume a chair conformation it is important to be aware of the possibility of conformational deformations as these may exert considerable influence.

Cycloheptanes and cyclooctanes

Unlike cyclohexane *106*, cycloheptane *130* and cyclooctane *131* do not have well defined minimum energy conformations. Rather,

130

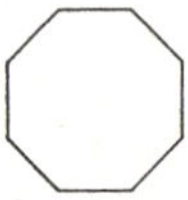

131

they have a number of flexible conformations (pseudorotation) which possess no angle strain, but variable torsional and van der Waals strains. The chair *132* and boat *133* are important conformations of cycloheptane *130* and twist forms of each are respectively of lower energy since relief of the torsional strain at C-4, C-5 and the trans-annular van der Waals strain between hydrogen atoms at C-3 and

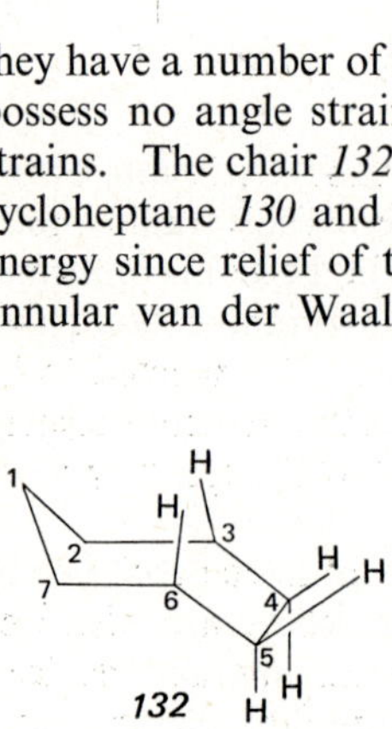

132

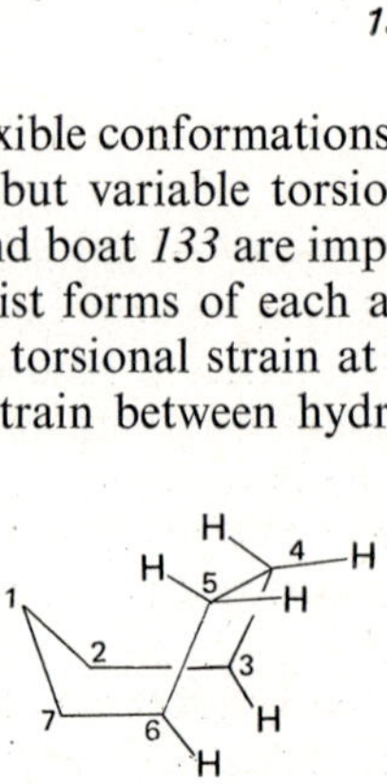

133

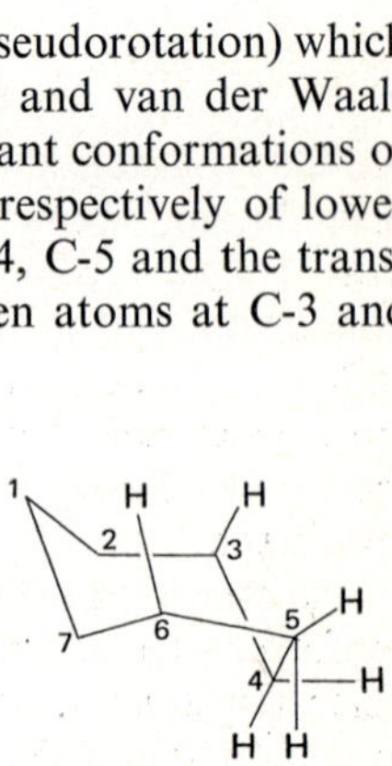

134

C-6 is afforded. In cycloheptane *130* itself the twist-chair conformation *134* appears to be the most stable whereas no uniform picture emerges for substituted cycloheptanes. The chair-boat *135* and the crown *136* are important conformations of cyclooctane *131* and the former appears to be preferred though in both there is torsional and trans-annular van der Waals strain (*eg* see hydrogen atoms on C-1 and C-5 in *135*).

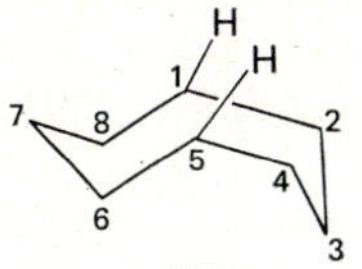

135

136

Conformations of fused carbocyclic systems

Decalins

Two fused cyclohexane chair conformers are involved in the stable conformers *138* and *140* of *trans*-decalin *137* and its *cis*-isomer *139* respectively. The ring fusion in *138* occurs through equatorial bonds (heavy lines) whereas that in *140* occurs through an equatorial and an axial bond. The consequence of this is that *trans*-decalin *137* and

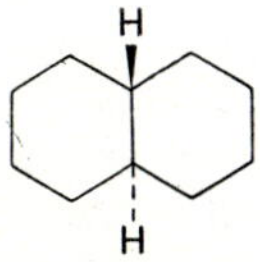

137

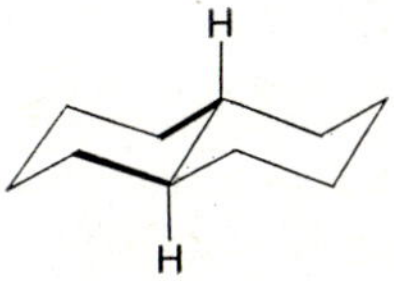

138

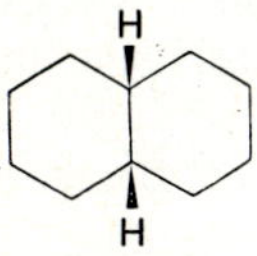

139

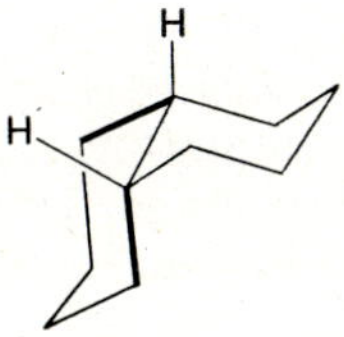

140

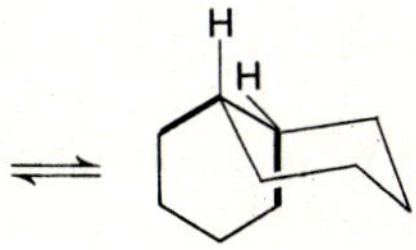

141

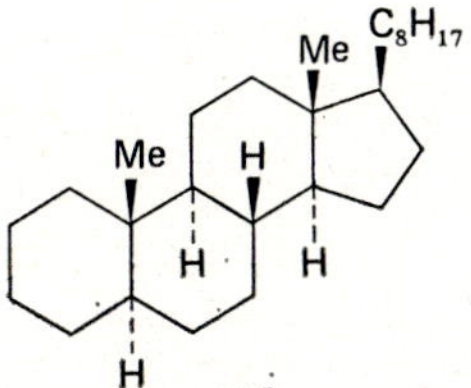

142

molecules containing this unit such as 5α-cholestane *142* are unable to undergo ring inversion and are thus conformationally fixed.* Such a ring inversion, which would involve conversion of equatorial to axial bonds and *vice versa*, would require one six-membered ring to be fused to the other through two *trans*-diaxial bonds, a requirement which is geometrically impossible. In contrast, *cis*-decalin *139* may undergo ring inversion as the ring junction which involves axial and equatorial bonds would in the ring inverted conformer simply involve equatorial and axial bonds respectively (*see 141*). The enthalpy of *cis*-decalin *139* is greater than that of *trans*-decalin *140* owing to its greater van der Waals strain and the experimentally observed value (*ca* 11.5 kJ mol^{-1}) is close to that predicted on the basis of the three additional *gauche* butane interactions (*ca* 10 kJ mol^{-1}) present in the former. Obviously, the introduction of substituents into the decalins could markedly influence the difference in stability of the *cis*- and *trans*-derivatives.

Hydrindanes

Owing to the near planarity of the cyclopentane ring, the fusion of such a ring with a cyclohexane ring will necessarily involve an increase in torsional strain as the bonds forming the ring junction will move towards an eclipsed conformation. This deformation is more easily accommodated for axial, equatorial bonds than for equatorial, equatorial bonds. Thus, *cis*-hydrindane *143* has only a slightly higher enthalpy (3.1 kJ mol^{-1} for liquid at 25 °C) than the *trans*-isomer *144*. The *trans*-isomer *144* is thermodynamically more stable than the *cis*-isomer *143* at 25° though, owing to the significantly larger standard entropy of the latter, which arises from its greater flexibility, the difference in standard free energy is small (*ca* 1 kJ mol^{-1}). As a result, at temperatures above 200 °C, the *cis*-isomer becomes the more stable. The relative stability of *cis*- and *trans*-isomers in hydrindanes in general is markedly influenced by substituents. For example, *cis*-hydrindan-1-one *145* is considerably more stable than the *trans*-isomer. It appears that generally the cyclohexane ring in hydrindanes adopts a slightly distorted chair conformation, but the precise

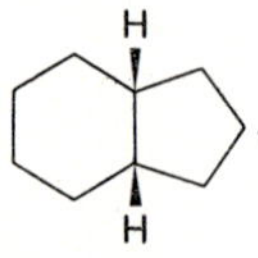

143

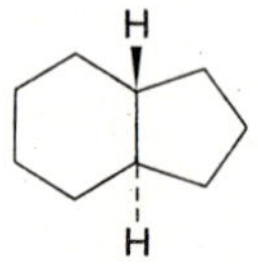

144

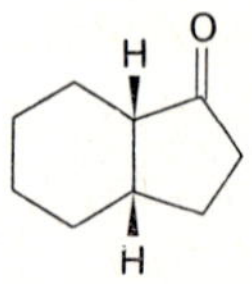

145

* The term conformationally fixed does not imply that small deformations or even deformations to the flexible form are impossible.

conformation of the cyclopentane ring is less certain. As in the decalins, the *cis*-isomers may undergo ring inversion while the *trans*-isomers may not and the energy barrier to ring inversion for the *cis*-hydrindanes is lower than that for the *cis*-decalins.

Conformations of selected heterocycles

As in the saturated carbocyclic compounds the conformations of six-membered heterocycles are better defined than those of five-membered heterocycles and a few examples of the former only will be discussed.

Tetrahydropyrans

By analogy with cyclohexane *106*, the chair conformer *147* of tetrahydropyran *146* would be expected to be more stable than the flexible conformer exemplified by *148*. This is found to be so and many

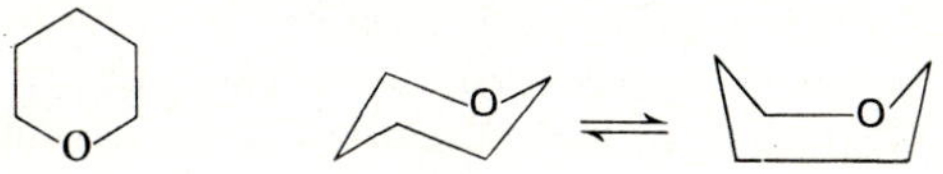

146 *147* *148*

carbohydrates, in particular, have been shown to adopt the chair conformation. For example, D-(+)-glucose *78* (*see Chapter 1*) in aqueous solution is essentially a mixture of the hemiacetals *149* and *150* (called β- and α-anomers respectively) in which all substituents, other than the anomeric hydroxyl group at C-l, are equatorial.

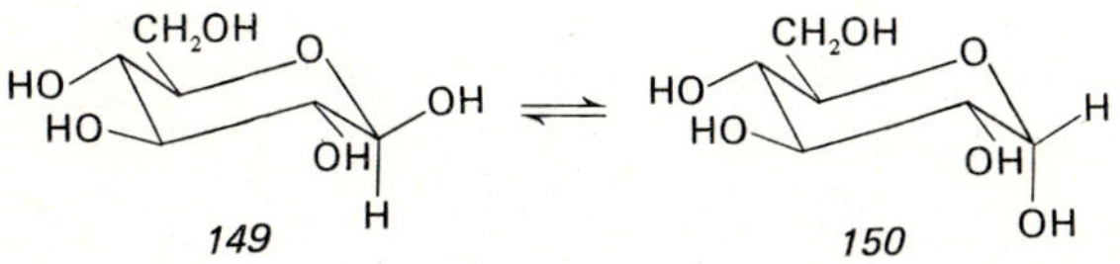

149 *150*

The β-anomer *149* which interconverts with the α-anomer *150 via* the open-chain form *78*, is the more stable in aqueous solution (36 per cent α:64 per cent β), but not as stable as would be predicted from consideration of *syn*-1,3-diaxial hydrogen interactions with the axial hydroxyl group. The reduced stability of the β-anomer *149* is a result of unfavourable interactions between the dipoles of the carbon–oxygen ring bonds and the C-1—OH bond (the so-called anomeric effect).

Piperidines

It would be expected that piperidine *151* would preferentially adopt a chair conformer. This is generally agreed but there is some controversy as to which of the two possible chair conformers *153* and *154* is favoured. No clear picture emerges, but Katritzky and co-workers conclude that the conformer *155*, in which the N—H bond is

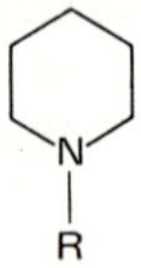

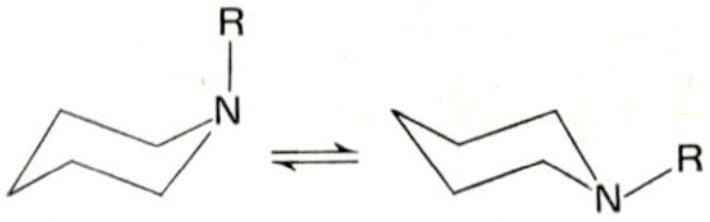

151 R = H
152 R = Me

153 R = H
154 R = Me

155 R = H
156 R = Me

equatorial, is favoured* in non-polar solvents. There appears to be general agreement that N–methylpiperidine *152* prefers to exist in equatorial conformer *156*. However, the precise energy difference between *156* and *154* is not entirely clear.

It should be stressed here that, as in the case of cyclohexane derivatives, substitution of the six-membered heterocyclic rings may cause distortion from the normal chair conformer and could lead to the flexible conformer being preferred.

* This view is supported by recent work by F. A. L. Anet and I. Yavari, *J. Am. Chem. Soc.*, 1977, **99**, 2794.

3. Physical Methods of Conformational Analysis

In the previous chapter, the conformational preferences of a number of simple molecules and the factors responsible for these preferences were discussed. Little of the experimental evidence which supports the existence of these preferred conformations was presented. Now examples of the applications of some of the physical methods employed in conformational investigations are presented. It will be assumed for the most part that the basic principles of each method are understood.

X-ray and electron diffraction

The diffraction patterns of X-rays from crystalline solids allow the total structures of the molecules to be determined. In molecules in which several conformers are possible, a particular crystalline form will generally contain a single conformer. For example, 1,2-dichloroethane crystallises exclusively in the *anti*-conformer *97* (*see Chapter 2*). Although this conformer is the thermodynamically most stable conformation, it may not be assumed that a crystalline lattice will necessarily always incorporate such a conformation as the stability of the lattice itself may have profound effects. In certain cases, dependent upon crystallising conditions, different crystalline forms incorporating different conformers may be isolated. *Eg* 1,1,2,2-tetrabromoethane *157* crystallises in the *gauche*-conformer *158* and the *anti*-conformer *159*. Accordingly, correlation of the behaviour of a

$Br_2CH{-}CHBr_2$

157

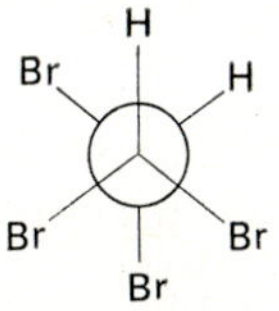

158

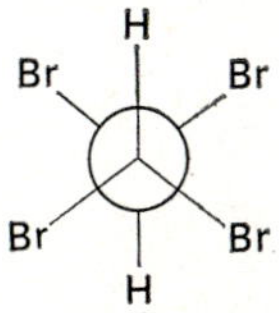

159

molecule in solution with its conformation in the solid state is not always meaningful and since most chemists are concerned with the former it is apparent that the utility of this method is limited.

Analysis of the electron diffraction patterns — usually of gaseous samples — provides all interatomic distances and, thereby, the total structure of the molecules. For a gaseous sample containing a number of conformers in equilibrium, a set of interatomic distances for each conformer may be determined. In addition, estimates of the populations of the conformers and conformational free energies may be deduced. The method is suitable only for relatively simple molecules. In an early investigation of 1,2-dichloroethane it was shown that the *anti*-conformer *97* was preferred to the *gauche*-conformer *99* (*see Chapter 2*) and it was estimated at 22°C that the distribution was *ca* 3:1. The presence of two conformers was indicated by two different Cl–Cl distances (*Fig. 18* ↔) and different Cl–H distances for chlorine and hydrogen bonded to different carbon atoms (*Fig. 18* ⇠⇢). In fact the *gauche*-conformer does not have the ideal staggered conformation indicated and exhibits two different *gauche*–H–Cl distances.

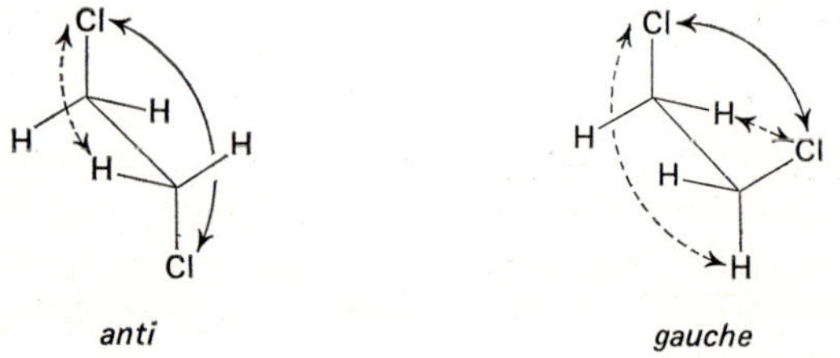

Fig. 18. Interatomic distances in 1,2-dichloroethane.

Electron diffraction demonstrated that 1,1,2,2-tetrabromoethane *157* is predominantly in the *gauche*-conformer *158* at 105 °C and that the cyclohexane ring, even in its unsubstituted form, is slightly flattened such that the C–C–C angles are slightly larger than tetrahedral, being 111.5°.

Estimates of the energy barriers to rotation about single bonds may be obtained using electron diffraction and vibrational motion studies.

As electron diffraction measurements are determined for gaseous samples, intermolecular forces are virtually absent and conclusions drawn from these studies may be quite different from those arrived at by other methods in the condensed phase.

Spectroscopic methods

Nuclear magnetic resonance spectra

Studies of conformational equilibria have been greatly facilitated by using chemical shift and spin–spin coupling constant data. Most of the early work in this area utilised ^{1}H nmr spectra, but more recently, with the availability of Fourier Transform nmr spectrometers, ^{13}C nmr spectra have also been utilised. Irrespective of the nucleus studied the same principles apply and the methods used are best described by consideration of some simple examples.

Consider the conformational equilibrium of a monohalogenated cyclohexane (*160*⇌*161*). The ^{1}H nmr spectra of *160* and *161* would be expected to be different and if the interconversion of *160* and *161* were slow on the nmr time scale, different signals for the equatorial proton (H_E) and the axial proton (H_A) would be clearly visible. The general trend of chemical shifts would suggest that the signal for H_A would be at higher field than that for H_E and, owing to the electronegative halogen, both signals would be at lower field than the remainder of the protons.

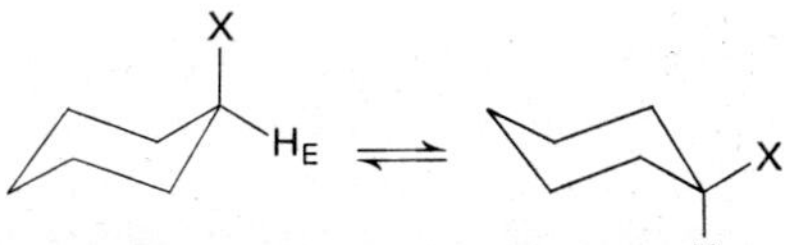

160 161

Integration of the signals for H_E and H_A would give directly the equilibrium concentrations of *160* and *161* and thus the equilibrium constant (K) and the standard conformational free energy ($\Delta G^\ominus = -2.303\ RT \log_{10} K$). Of course, at room temperature the interconversion of *160* and *161* is very rapid and the ^{1}H nmr spectrometer records an averaged spectrum. However, at *ca* $-100°$C the rate of interconversion is sufficiently slow for both conformers to exhibit discrete spectra and thus the conformational free energy may be determined (*eg* when X=Br, $-\Delta G^\ominus = 2$ kJ mol^{-1}). The value thus determined relates only to the particular temperature of measurement and although for many monosubstituted cyclohexanes this will be close to the room temperature value owing to $\Delta S^\ominus$ being small ($\Delta G^\ominus = \Delta H^\ominus - T\Delta S^\ominus$) this condition will not always apply. A further limitation on low temperature studies arises for equilibria in which the less stable conformer is present only to the extent of say 5 or 10 per cent at room temperature. On cooling, the less stable conformer may become undetectable owing to the increase in the equilibrium constant favouring the more stable conformer.

The room temperature equilibrium constant and $\Delta G^\ominus$ may be determined from the low temperature chemical shifts of H_E and H_A and the average chemical shift (δ_{Av}) observed at room temperature provided the former are corrected for their temperature dependence. The following equation relates the mole fractions, x and $1-x$, of the two conformers, the average chemical shift (δ_{Av}) and the chemical shifts of H_A (δ_A) and H_E (δ_E) at any given temperature.

$$\delta_{Av} = x\delta_A + (1-x)\delta_E$$

An alternative method of determining δ_A and δ_E at room temperature is to measure the chemical shifts of *cis*- and *trans*-4-*t*-butylcyclohexanes. The assumption here is that in the *cis*-model compound, which will exist essentially in the conformation *162* in which

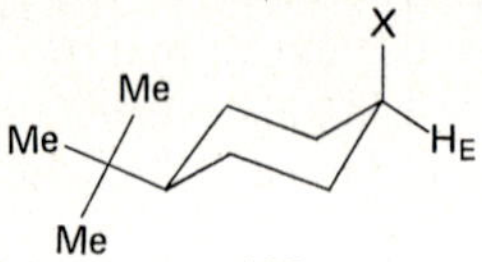

162

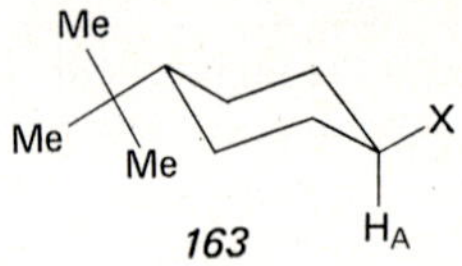

163

the *t*-butyl group is equatorial, the chemical shift of H_E will be identical to that in *160*. Similarly H_A in *163* is assumed to have the same chemical shift as H_A in *161*. This assumption is not entirely valid as the *t*-butyl group influences the chemical shifts and only approximate $\Delta G^{\ominus}$ values may be obtained by this method.

A further application of variable temperature nmr spectroscopy is in the determination of rotational free energy barriers. For example, N,N-dimethylacetamide *164* has restricted rotation about the N—CO bond, owing to partial double bond character in that bond, and thus has a planar conformation which interconverts slowly with an equivalent planar conformation at room temperature.

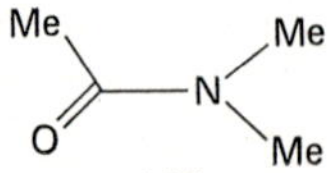

164

The ^1H nmr spectrum at room temperature exhibits different singlets for the methyl groups attached to nitrogen, but as the temperature is increased these signals become broader and coalesce and finally at around 90 °C become a single sharp singlet. This transition corresponds to a gradual increase in the speed of rotation about the N—CO bond until the nmr signals for the N—Me groups become averaged. The free energy barrier to rotation (*ca* 76 kJ mol^{-1}) may be determined from the coalescence temperature (*ca 65*°C) and the difference in frequencies of the signals of the N—Me groups at room temperature. Similarly observation of the ^1H nmr spectrum of cyclohexane *106* over the temperature range −100 °C to room temperature establishes a coalescence temperature of *ca* −67 °C for the signals for the axial and equatorial protons and an energy barrier to ring inversion of *ca* 43 kJ mol^{-1}. In certain cases, particularly when the spectrum shows extensive spin–spin coupling, it is not possible to observe any significant temperature variation. The use of ^{13}C nmr spectroscopy may circumvent this problem since all

^{13}C–^{1}H couplings can be removed and there is no ^{13}C–^{13}C coupling. A recent study of *cis*-decalin *139* has employed this technique and established an energy barrier of *ca* 52 kJ mol^{-1} for the interconversion *140* ⇌ *141* (*see Chapter 2*).

The magnitude of spin–spin coupling constants (J) between nuclei on adjacent carbon atoms depends on the dihedral angle (ϕ), and thus may provide useful conformational data. The Karplus equation

$$J = a \cos^2 \phi - c$$

in which *a* and *c* are constants, expresses this dependence for vicinal protons. The value of *a* depends on the range of ϕ and when $0° \leqslant \phi \leqslant 90°$ the nominal value for *a* is 10 whereas when $90° \leqslant \phi \leqslant 180°$ it is 16. Other factors such as the electronegativity and orientation of substituents further influence the value of *a*. In general terms, values of J for *anti*-protons are greater than those for *gauche*-protons. Accordingly, it is easy to recognise that H_E in the conformationally biased halocyclohexane *162* is equatorial since the bandwidth at half-height* of its signal in the ^{1}H nmr spectrum is characteristic (*ca* 10 Hz) of four couplings with vicinal *gauche*-protons.

Conversely, the signal for H_A in *163* will have a larger (*ca* 25 Hz) bandwidth at half-height owing to the vicinal coupling with two *anti*-protons and two *gauche*-protons and H_A is recognised as being axial. In this particular example, the chair conformations of *162* and *163* are confirmed since their configurations are known. The same principles apply to rather more complex conformationally biased or rigid molecules and, in particular, distortions from the chair conformation have often been detected.

In conformationally mobile molecules the observed coupling constants will be the weighted average of the values associated with each conformer and provided these individual J values are known the equilibrium populations of the conformers present may be determined. In practise, this is often difficult to achieve accurately owing to the uncertainty in the J values determined from the Karplus equation or from conformationally biased or rigid models. Despite these difficulties there are many applications both in acyclic and alicyclic compounds. For example, the conformational free energy for 3,3,4,-4,5,5-hexadeuteriocyclohexanol *165* was calculated (*ca* 4.2 kJ mol^{-1}) by this method. The expected coupling constants associated with each chair conformer of *165* were obtained from the spectra of the conformationally biased models *166* and *167*. Incorporation of deuterium into substrates, as exemplified here, simplifies the spin–spin coupling patterns and this is quite a general technique.

* The bandwidth at half-height gives the total coupling when the individual coupling constants may not be determined owing to the lack of any resolved structure.

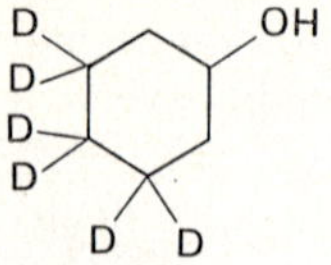
165

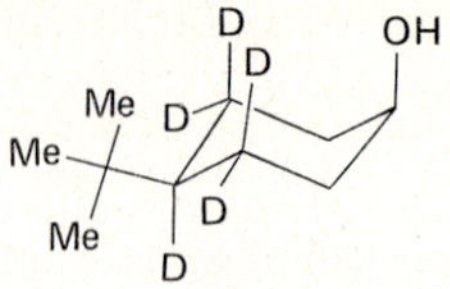
166

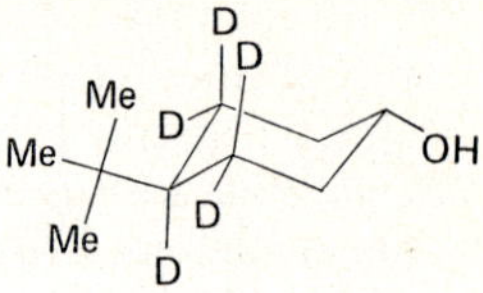
167

Lanthanide shift reagents, such as Eu $(dpm)_3$,* can also be used to simplify the analysis of nmr spectra, but the possibility of the shift reagent itself inducing conformational changes has to be considered when interpreting the data.

Infrared and Raman spectra

It is possible to use the characteristic absorption bands of certain functional groups to obtain conformational information provided their frequencies differ for different conformations. For example, for chlorocyclohexane the conformational equilibrium *160* ⇌ *161* (X=Cl) in the vapour state is indicated by the presence, in the ir spectrum, of the equatorial and axial C–Cl stretching bands at 742 and 688 cm^{-1} respectively.† In general, for two such conformers (1 and 2) the equilibrium constant (K) is given by

$$K = \frac{I_2 \alpha_1}{I_1 \alpha_2}$$

where I_1 and I_2 are the measured extinctions and α_1 and α_2 are the integrated molar extinctions for the bands being observed. Unfortunately the values of α are not known and values are estimated by use of conformationally rigid or biased models. For example, the ir spectra of (*162*, X=Cl) and (*163*, X=Cl) could be used to provide α for *160* and *161* respectively, thereby allowing K and the conformational free energy to be calculated.

The difference in standard enthalpy between conformers can be determined without knowledge of the α values. If I_1 and I_2 are determined over a range of temperatures, a plot of ln (I_2/I_1) against reciprocal temperature yields $\Delta H^{\ominus}$ since the following equation can be derived. The conformational free energy may be obtained since α_2/α_1 may be determined from the slope of a plot of I_2 against I_1 and thus $\Delta S^{\ominus}$ may be calculated.

$$\ln\left(\frac{I_2}{I_1}\right) = -\frac{\Delta H^{\ominus}}{RT} + \ln\frac{\alpha_2}{\alpha_1} + \frac{\Delta S^{\ominus}}{R}$$

* Eu $(dpm)_3$ is tris (dipivalomethanato) europium (III).

† In general for monosubstituted cyclohexanes the C–X stretching band for an equatorial substituent is at a higher frequency than that for an axial substituent.

Ir spectroscopy is particularly useful for measuring hydrogen bonding and is therefore useful for conformational studies of vicinal diols and similar molecules where intramolecular hydrogen bonding is important. The O–H stretching band for a hydrogen bonded OH appears at lower frequency (near 3590 cm^{-1}) than in an unbonded OH (near 3630 cm^{-1}) and the separation ($\Delta\nu$) of the bonds increases with the strength of the hydrogen bond. If one of a mixture of two conformers shows no hydrogen bonding while the other does then, using the principles described above, the conformational free energy or enthalpy could be determined from the extinctions of the bonded and non-bonded bands. The strength of the hydrogen bond, as determined from $\Delta\nu$, may provide additional information. For example, $\Delta\nu$ for *cis*-cyclohexane-1,2-diol *168* is 39 cm^{-1} while for the *trans*-isomer *169* it is 32 cm^{-1}. This indicates that the OH

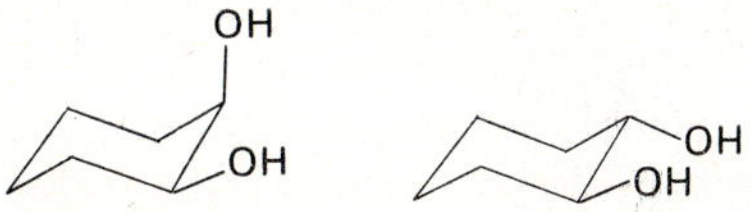

168 *169*

groups in *168*, which are axial, equatorial, are able to move closer together than in *169* where they are equatorial, equatorial. This greater resistance to deformation of diequatorial versus equatorial, axial bonds is a general feature of cyclohexane chemistry.

Ultraviolet spectra

Owing to the limited number of structures which absorb radiation in the uv region of the spectrum there are relatively few conformational problems which may be solved from a study of uv spectra.

Disruption of the conjugation of a chromophore by rotation around a single bond may be detected. As was shown (*see Chapter 1*) *ortho*-substituted biphenyls will not readily take up the conformation in which the aromatic rings are co-planar and fully conjugated and this is reflected in their uv spectra. For example, 3,3′-dimethylbiphenyl *170* has λ_{max} 255 nm (ϵ 14 000) whereas the isomeric 2,2′-dimethylbiphenyl *171* has a much lower intensity band (ϵ 800) at λ_{max} 260 nm confirming the loss of conjugation. Similar observations have been made for sterically hindered aryl aldehydes and ketones.

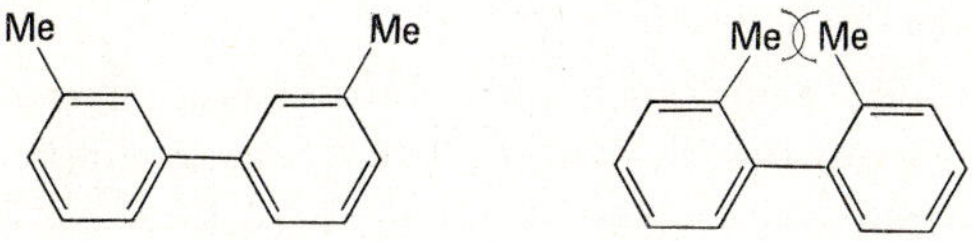

170 *171*

The low intensity n–π^* transition for cyclohexanones is shifted to longer wavelength by axial substituents on the α-carbon such as halogen, phenyl, and hydroxyl. The same equatorial substituents have little effect and it is possible therefore to obtain useful conformational information from the uv spectra of selected cyclohexanone derivatives.

Dipole moments

Since dipole moments are vectors, the total dipole moment of a molecule is often useful in providing conformational information. For a molecule which exists in two conformers A and B in equilibrium, their mole fractions — x and $1-x$ respectively — are related to the measured dipole μ and their individual dipoles — μ_A and μ_B — by the following equation.

$$\mu^2 = x\,\mu_A^2 + (1-x)\,\mu_B^2$$

Accordingly, it is a simple matter to calculate x, the equilibrium constant and the conformational free energy provided μ_A and μ_B may be estimated. This is usually achieved by calculation using individual group dipole moments or by the measurement of dipole moments of suitable models.

In the case of 1,2-dichloroethane (*Fig. 18*), the expected dipole moment of the *anti*-conformer would be zero while that of the *gauche*-conformer would be *ca* 3.2 D. Gaseous 1,2-dichloroethane has a dipole moment of 1.12 at 32°C and 1.54 at 270°C indicating that 88 per cent of the molecules are in the *anti*-conformer at the lower temperature and this figure is reduced to 77 per cent at the higher temperature owing to the favourable entropy of the *gauche*-conformer.

The dipole moment of *trans*-1,2-dibromocyclohexane *125* (*Chapter 2*) in carbon tetrachloride at 30°C is 1.76 D. Assuming that the dipole moment of the diaxial conformer *127* is zero and that of the diequatorial conformer *126* is the same as the *cis*-1,2-dibromocyclohexane *172* (3.13 D) it may be calculated that 32 per cent of the molecules are in the diequatorial conformation.

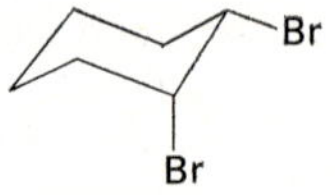

172

Conformational equilibria in polar molecules are markedly affected by solvent polarity and, in benzene, the observed dipole moment of *trans*-1,2-dibromocyclohexane increases to 2.16 D indicating that 48 per cent of the molecules are in the diequatorial conformer.

Benzene is relatively more polar than carbon tetrachloride and stabilises the diequatorial conformer *126* more than the diaxial conformer *127* owing to the greater dipole moment of the former.

Comparison of calculated and experimentally determined dipole moments in conformationally rigid or biased molecules provides insight into the general geometry of the molecules and in particular whether distortions from the expected conformations are present. For example, dipole moment measurements indicate that, in solution, the A-ring of 4,4-dimethyl-5α-androstane-3,7-dione *173* has a flattened chair conformation *174*. The flattening, which arises from the diaxial interaction between the β-methyl groups at C-10 and C-4 is confirmed in the solid state of other 4,4-dimethyl-3-oxo-5α-steroids by X-ray crystallographic studies.*

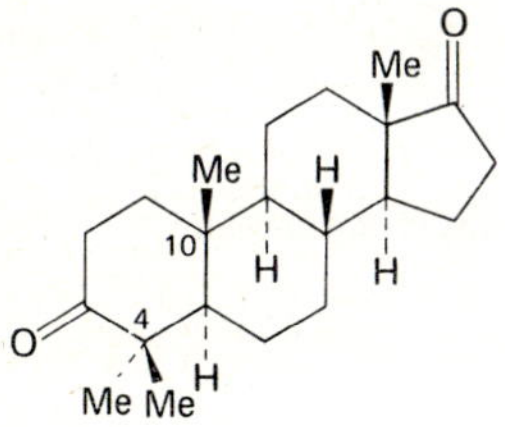

173

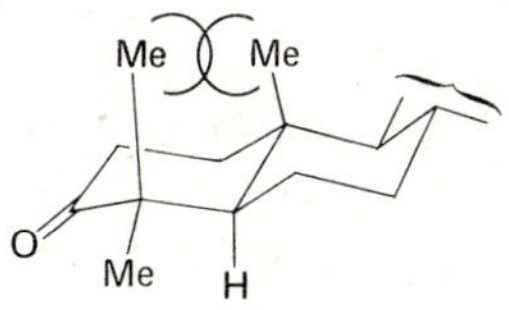

174

It is difficult to obtain precise conformational data from dipole moment measurements particularly of complex molecules owing to the inherent inaccuracies in the determination of the expected dipole moments. However, the method has been widely used and it can provide very meaningful data particularly if it is applied in conjunction with other methods.

Chiroptical techniques: optical rotatory dispersion and circular dichroism

The determination of the specific rotation (*see Chapter 1*) of optically active molecules is conventionally carried out at the wavelength of the sodium D-line (589 nm). However, the specific rotation is found to vary with the wavelength of light used and the dependence of molecular rotation ($[\alpha]^t_\lambda \times$ molecular weight/100) on wavelength is termed **optical rotatory dispersion** (ord). The ord curve may be plain or it may exhibit a Cotton effect (*Fig. 19*).

A plain curve is obtained for optically active molecules which do not have a chromophore that absorbs light in the spectral region

* See U. Burkert and N. L. Allinger, *Tetrahedron*, 1978, **34**, 807 and J. M. Midgley, J. E. Parkin, and W. B. Whalley, *J. Chem. Soc., Perkin Trans. I*, 1977, 834 for full discussion.

under study whereas the Cotton effect curve is obtained for molecules processing such a chromophore. Molecules exhibiting a Cotton effect in the ord curve also exhibit circular dichroism (cd). The cd curve shows the dependence of molecular ellipticity on wavelength (*Fig. 20*) and the sign of the Cotton effect is the same as that in the ord curve. The molecular ellipticity is determined from the differential absorption of the left- and right-circularly polarised components

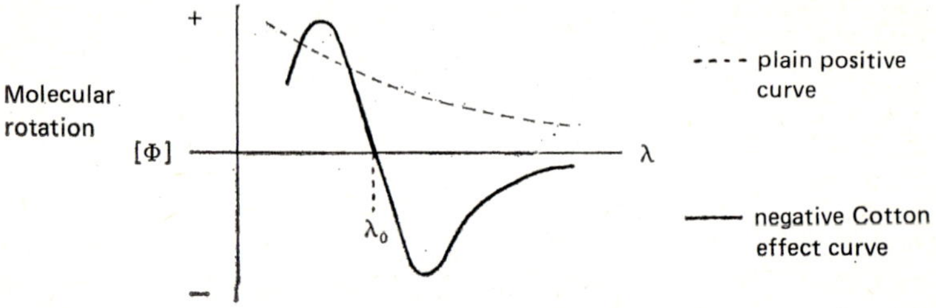

Fig. 19. Optical rotatory dispersion curves.

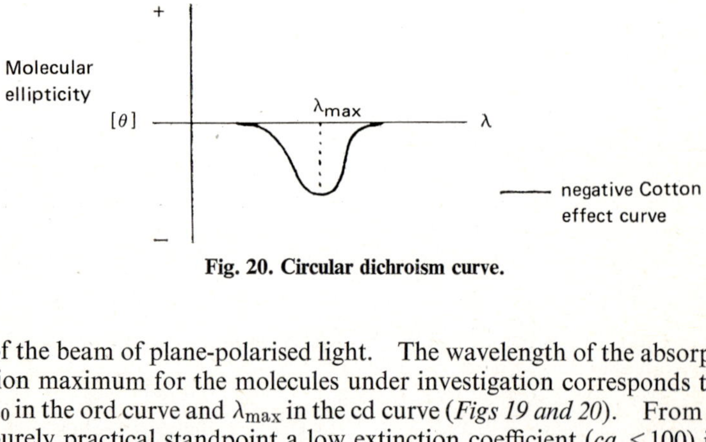

Fig. 20. Circular dichroism curve.

of the beam of plane-polarised light. The wavelength of the absorption maximum for the molecules under investigation corresponds to λ_0 in the ord curve and λ_{max} in the cd curve (*Figs 19 and 20*). From a purely practical standpoint a low extinction coefficient (*ca* $\leqslant 100$) is required. The sign and amplitude of the Cotton effect in both ord and cd curves may be used to determine conformation provided the absolute configuration is known.

Cyclohexanones, which exhibit a low intensity absorption band (n–π^*) at 290–300 nm, have perhaps been the most widely studied and have led to the formulation of the octant rule which allows the sign and amplitude of the Cotton effect to be predicted.

The cyclohexanone is divided into octants using three planes X, Y, and Z (*Fig. 21*). The C=O bond is bisected by the X-plane and lies in the Y and Z planes and C-2 and C-6 lie in the Y plane while C-4

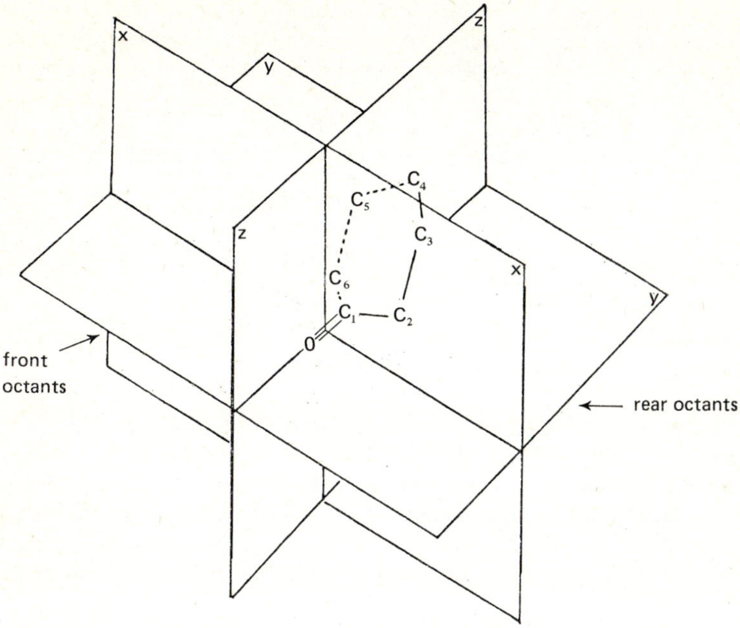

Fig. 21. Octant rule for cyclohexanone.

lies in the Z plane. There are thus four rear octants which are on the opposite side of the X plane from the oxygen atom, and which generally contain the substituent groups or atoms, and four front octants which are on the same side of the X plane as the oxygen atom. Substituents lying in planes Y and Z make no contribution to the Cotton effect and this includes substituents at C-4 and equatorial substituents at C-2 and C-6. The signs of the contributions of other substituents depend on the particular octant in which they lie (*Fig. 22*) and the sizes of the contributions depend on the nature and position of the substituents.

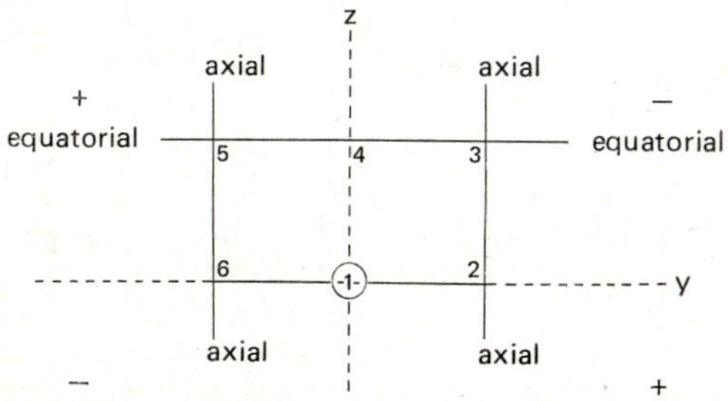

Fig. 22. Projection showing signs of contributions of substituents in rear octants.

In particular, axial groups at C-2 and C-6 make major contributions and values for a series of substituents have been assigned thus allowing reasonably accurate quantitative predictions of the Cotton effect.

From the application of the octant rule to (+)-*trans*-10-methyl-2-decalone *175*, the observed positive Cotton effect would be predicted

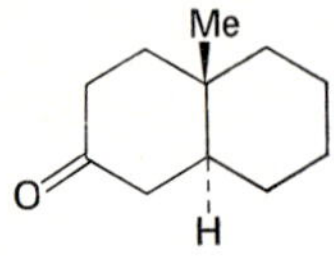

175

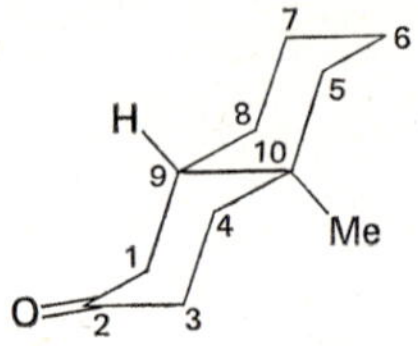

176

on the assumption that the two chair conformation *176* is involved since C-6, C-7 and C-8 each make a positive contribution and there are no negative contributions (*Fig. 23*). Effectively the conformation *176* is confirmed. Although this is not particularly surprising ord/cd studies have uncovered some quite unpredictable

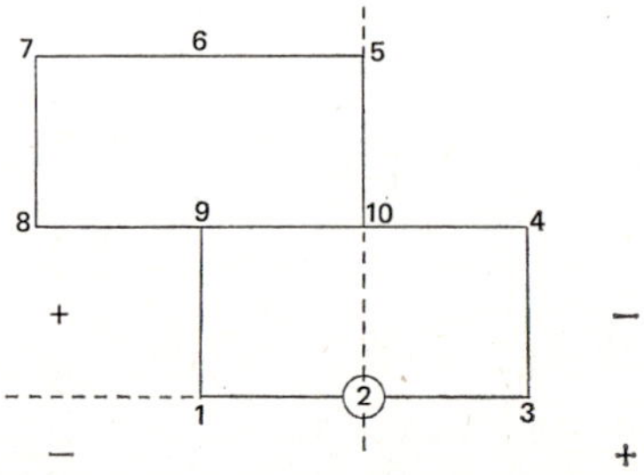

Fig. 23. Octant rule applied to (+)-trans-10-methyl-2-decalone.

conformational preferences. For example, a negative Cotton effect for 17β-acetoxy-2α-bromo-2β-methyl-5α-androstan-3-one *177* indicates that the A-ring boat conformer *178* is adopted,* and (+) isomenthone *179*, having a positive Cotton effect, was shown to exist mainly in the conformer *180* which possesses the axial isopropyl (1-methylethyl) group. The latter result is of course only qualitative since (+) isomenthone *179* is a conformationally mobile system. In conformationally rigid or biased molecules the ord/cd curves represent a single conformation whereas in conformationally mobile

* Conversion into a boat form in cyclohexanones is much less energetically demanding than in cyclohexanes.

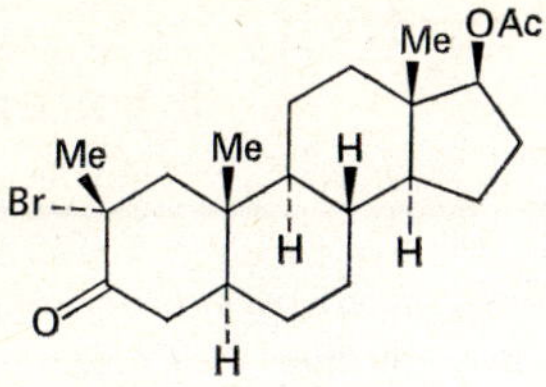

177

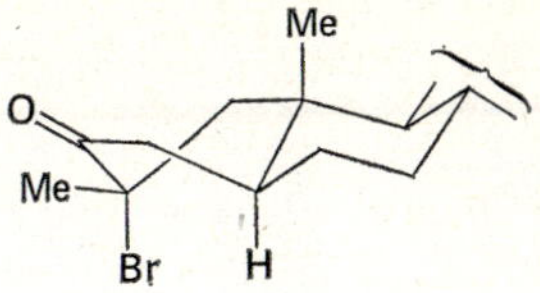

178

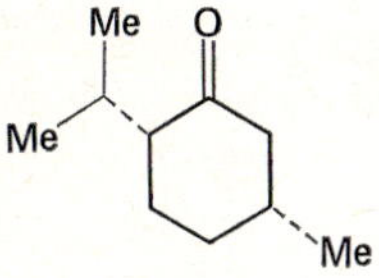

179

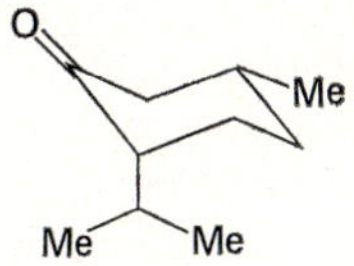

180

systems each conformer shows a band and the resultant curve is a composite of these, overlapping, bands. Studies of the variation of the composite cd curve with temperature can, in certain cases, provide reasonable values of equilibrium constants.

4. Conformation and Reactivity in Cyclic Systems

In general terms, reaction rates may be influenced by van der Waals, torsional, and bond angle strains since any changes in these *steric** factors involved in the conversion of reactants in their ground states to their transition states will obviously influence the enthalpy and the free energy of activation.† In reactions where two products may be formed such factors may influence the product ratio and in reversible processes they may influence the equilibrium constant. Additionally certain reactions are subject to *stereoelectronic** factors since the transition states have specific geometrical requirements for partly formed bonds. Thus the E2 reaction (bimolecular elimination) proceeds most efficiently when the eliminated groups have the *anti*-conformation (*Fig. 24*) and any departure from this will increase the energy of the transition state.

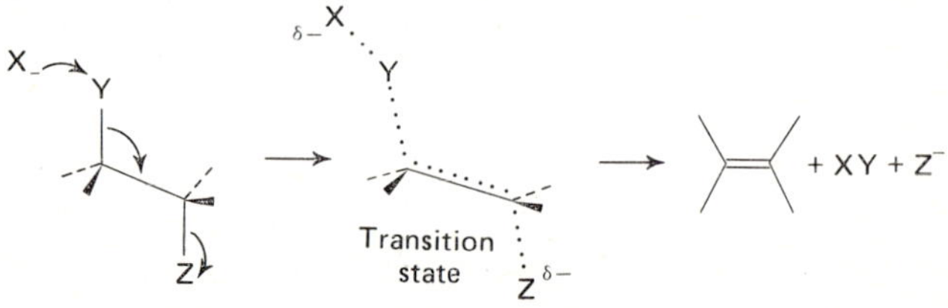

Fig. 24. E2 reaction.

Cyclohexane and fused cyclohexanes

The influence of conformation on reactivity is most easily studied in cyclohexanes and fused cyclohexanes owing to their well defined conformations. The energetically favourable chair conformer is most often involved and, in the *trans*-decalins and similar steroids and triterpenes, ring inversion is precluded. In simple cyclohexane derivatives reactions may be assumed to involve a single chair conformer, provided a sufficiently large 'anchoring' substituent (*eg* *t*-butyl) is present. The situation contrasts with that in acyclic

* The terms *steric* and *stereoelectronic* factors are discussed by J. McKenna, *Conformational analysis of organic compounds*, RIC Lecture Series, 1966, No 1, p 11. The former is used in its broadest sense and does not imply only the consideration of steric hindrance (van der Waals repulsion).

† From transition state theory, the rate constant $k_{obs} = \kappa kT/h \exp(-\Delta G^{\ominus\ddagger}/RT)$ where κ is the transmission coefficient and k and h are Boltzmann's constant and Planck's constant respectively, and the standard free energy of activation $\Delta G^{\ominus\ddagger} = \Delta H^{\ominus\ddagger} - T\Delta S^{\ominus\ddagger}$ where $\Delta H^{\ominus\ddagger}$ and $\Delta S^{\ominus\ddagger}$ are the standard enthalpy and standard entropy of activation. Except in special circumstances, $\Delta G^{\ominus\ddagger} \approx \Delta H^{\ominus\ddagger}$ since $T\Delta S^{\ominus\ddagger}$ is small. (*See also p 32.*)

molecules where conformational preferences are less well defined (*see Chapter 5*).

The differences in reactivity of axial versus equatorial substituents may be demonstrated by consideration of some common reactions.

Hydrolysis of esters and esterification of alcohols

The rate of hydrolysis of ethyl *cis*-4-*t*-butylcyclohexanecarboxylate *181* in aqueous ethanolic sodium hydroxide at 25 °C is around 20 times slower than that for the *trans*-isomer *182* owing to the carboethoxy group being axial in the former and equatorial in the latter.

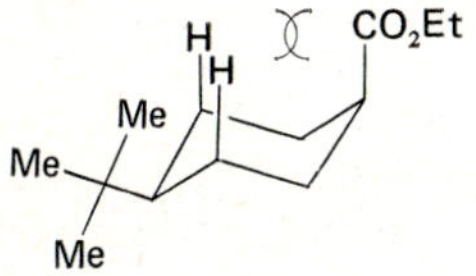

181

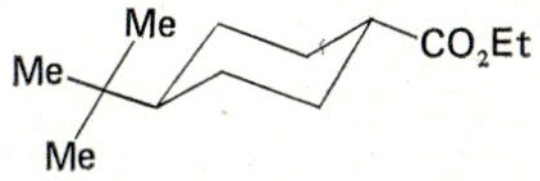

182

In general axial groups are more sterically hindered than equatorial ones (*see Chapter 2*) and approach by any reagent to the axial groups is correspondingly more hindered. Since the rate determining step in base-catalysed ester hydrolysis is the attack by the base at the carbonyl group it follows that *181* would be expected to react more slowly than *182*.

Similarly the rate of acetylation of *trans*-4-*t*-butylcyclohexanol *122* in acetic anhydride and pyridine at 25 °C is *ca* four times that of the *cis*-isomer *123* (*see Chapter 2*).

The ratio of the rates of hydrolysis of *trans*- and *cis*-4-*t*-butylcyclohexyl acetates *183* and *184* respectively in aqueous dioxan at 40 °C is *ca* seven, and further demonstrates the difference in reactivity of axial and equatorial substituents. It is noticeable that the difference in reactivity of *183* and *184* is not as pronounced as that of *181* and *182*. At first sight, this would be readily predicted since the carbonyl group in *184* which is attacked by the base is one bond further removed from the cyclohexane ring than in *181* and the effect of the *syn*-1,3-diaxial hydrogen atoms would be expected to be less marked. This is perhaps a facile explanation, however, since the site of reaction can, by rotation around the ring carbon–oxygen bond, be so far removed from the influence of the cyclohexane ring that little difference in reactivity would be expected for the axial and equatorial ester groups. It has been suggested* that the observed difference for cyclohexyl esters rather arises from steric hindrance to

* See P. B. Smith and J. McKenna, *Tetrahedron*, 1964, **20**, 1933 for further discussion.

solvation of the axial *vs* equatorial oxygen atom which acquires negative charge during the rate determining step of the reaction (*see 185*).

183

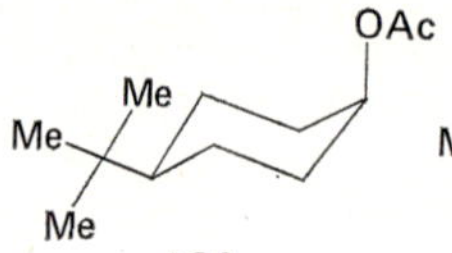

184

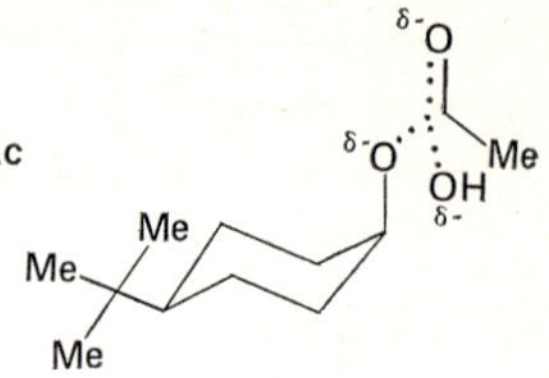

185

Not all axial cyclohexyl esters are hydrolysed more slowly than the corresponding equatorial compounds. For example, 3α-acetoxy-5α-cholestan-5-ol *186* is hydrolysed more readily than the 3β-epimer *187* owing to internal hydrogen bonding which stabilises the transition state for the former.

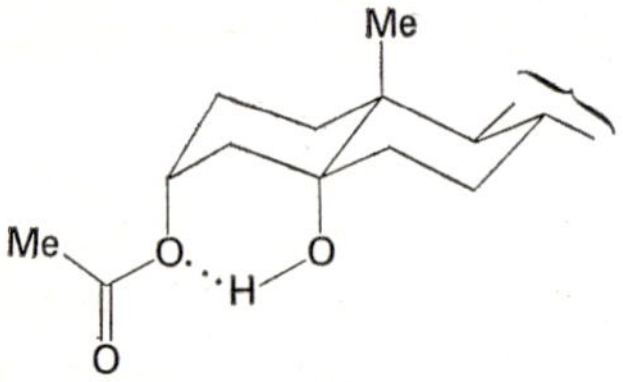

186

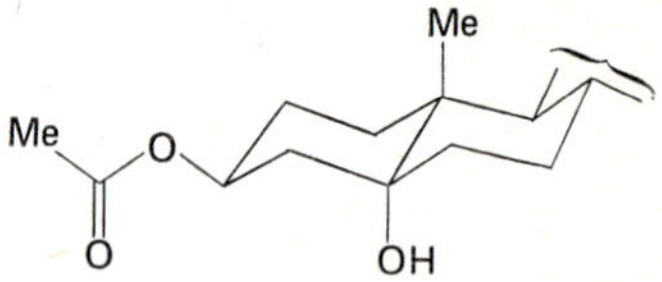

187

This example illustrates the general point that in considering conformation and reactivity the presence of polar groups may lead to a result which would not be predicted from a consideration of steric-factors alone (*see also Chapter 2*).

S_N1 and E1 reactions

The rate determining step in these unimolecular nucleophilic substitution (S_N1) and elimination (E1) reactions involves heterolysis of the bond between a carbon atom and a good leaving group (X) leading to an intermediate carbocation which may react with a nucleophile or may lose a proton (*Fig. 25*).

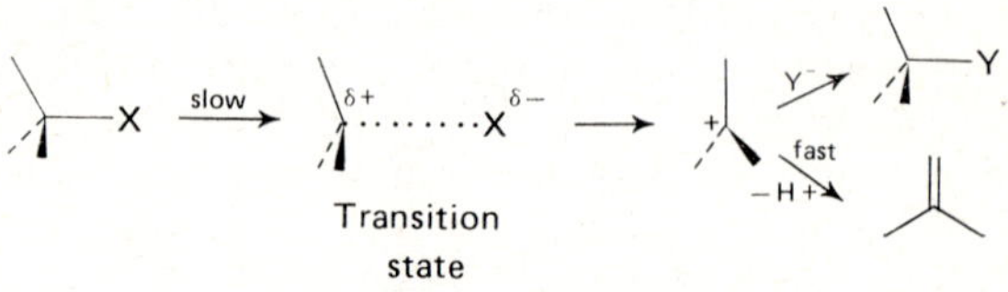

Fig. 25. E1/S_N1 reactions.

In general axial groups X react more quickly than the corresponding equatorial ones. For example, various solvolyses of *cis*- and *trans*-4-*t*-butylcyclohexyl tosylates *188* and *189* show that the *cis*-isomer *188* consistently reacts more rapidly than the *trans*-isomer

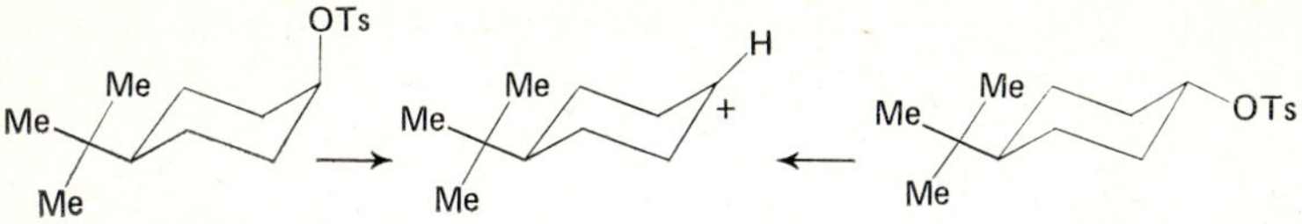

188 *190* *189*

189. Owing to the steric interactions between the axial tosyloxy group and the *syn*-axial hydrogen atoms, the *cis*-isomer *188* has a higher ground state energy than the *trans*-isomer *189*. The carbocation *190* may, as a first approximation, be assumed to be common to the reactions of both *188* and *189* and the transition states from both of these may also be assumed to have approximately the same energy. Accordingly the standard free energy of activation for the *cis*-isomer *188* will be less than that for the *trans*-isomer *189* and the former will react more quickly (*Fig. 26*). The steric compression suffered in the *cis*-isomer *188* is thus relieved in the transition state and the increased rate observed for this *vs* the *trans*-isomer is said to be a result of **steric acceleration**. It should be stressed that *Fig. 26* is only approximately representative as, in detail, the transition states and the intermediate carbocations produced from *188* and

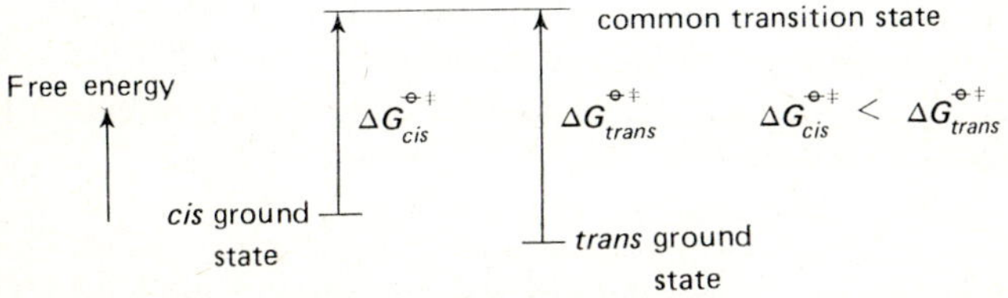

Fig. 26. Steric acceleration.

189 are not identical: different ratios of substitution *vs* elimination products are obtained and the substitution products do not have the same stereochemistry. Furthermore there is evidence that the *trans*-isomer *189* solvolyses, at least in part, in the flexible rather than the chair conformation.

Chromic acid oxidation of cyclohexanols to cyclohexanones

The rate determining step in this reaction is the decomposition of the rapidly formed chromate ester. Accordingly, owing to steric acceleration, axial hydroxy groups are oxidised more rapidly than the corresponding equatorial groups. Thus, *trans*-4-*t*-butylcyclohexanol *122* is oxidised at 25°C approximately four times more slowly than the *cis*-isomer *123* (*Chapter 2*). The transition states involved in converting the chromate ester into the common product, 4-*t*-butylcyclohexanone *191*, may be assumed to have similar energies and *Fig. 26* approximately represents the energy changes involved.

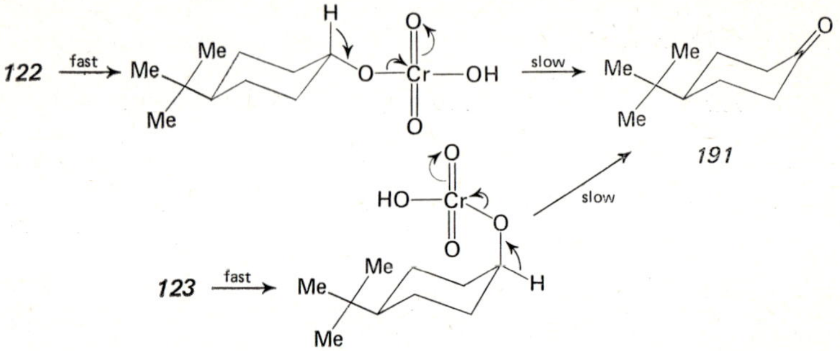

Hydride reductions of cyclohexanones to cyclohexanols

The factors controlling the stereochemical outcome of the reductions of cyclohexanones to cyclohexanols, using lithium aluminium hydride, sodium borohydride and similar reagents are still not fully understood.* A number of important features do, however, emerge. In general, ketones that are sterically hindered by axial substituents at the 3- and 5-carbon atoms give preferentially the axial alcohol owing to preferential approach of the reagent from the least hindered side (**steric approach control**). The proportion of axial alcohol in the product is influenced by the number and size of the axial substituents and the size of the reagent molecule. The latter may be markedly influenced by solvation. The alternative axial mode of attack leading to the equatorial alcohol usually occurs preferentially in ketones having no 3,5-axial substituents. Thus, whereas 3,3,5-trimethylcyclohexanone *192* gives 55 per cent of the axial alcohol *193* with lithium aluminium hydride, 4-*t*-butylcyclohexanone *190* gives largely (90 per cent) *trans*-4-*t*-butylcyclohexanol *122*. It was suggested by Dauben, Noyce and co-workers that the stereochemistry of the latter reaction and similar reactions with unhindered ketones was a result of **product development control**. The view was that in the absence of any steric hindrance thermodynamic factors determined the course of the reaction. While this concept was useful in rationalising a number of results, an unsatisfactory implication is

* See D. M. S. Wheeler and M. M. Wheeler in *Organic reactions in steroid chemistry* (eds J. Fried and J. A. Edwards). New York: Van Nostrand Reinhold, 1972, Vol. 1. p 66.

that the transition states for the reductions of unhindered ketones are product-like while those for hindered ketones are reactant-like.

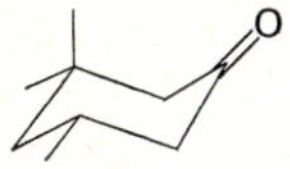

192

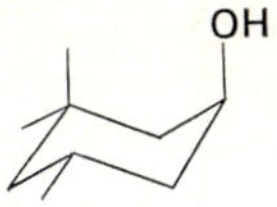

193

Felkin and his co-workers have offered an attractive alternative explanation for the stereochemical course of these reductions and the related reactions with Grignard reagents. It is proposed that the transition states for all reactions are reactant-like. The transition state *194*, leading to the equatorial product is subject to steric hindrance by the axial groups at C-3 and C-6 while the transition state *195* leading to the axial product is subject to torsional strain between the axial bonds at C-2 and C-6 and the partial bond between C-1 and the incoming nucleophile The use of H- in *194* and *195* is schematic and does not imply that the effective nucleophile is a simple hydride ion and the stereoelectronically controlled attack of the nucleophile

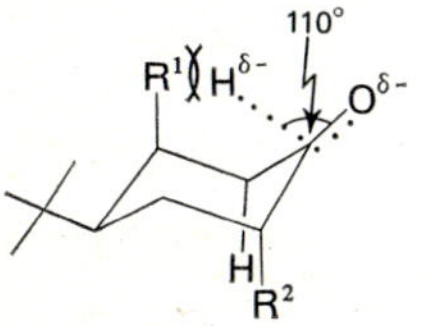

194

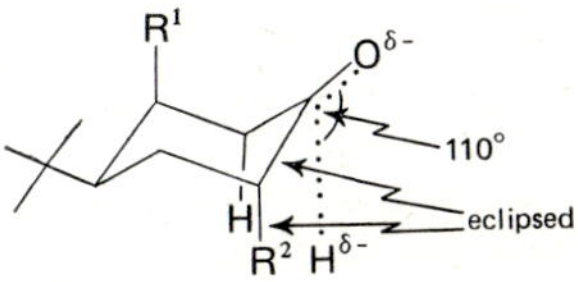

195

at an angle of *ca* 110° to the C—O bond is in accord with recent theoretical and experimental results. The stereochemical course of the hydride reductions may, from this viewpoint, be seen to depend on the degree of steric and torsional strain in the two alternative transition states and the concept of product development control is not required.*

E2 reactions

As indicated in the introduction to this chapter, these bimolecular elimination reactions are subject to stereoelectronic factors and proceed most efficiently when the eliminated groups have the *anti*-conformation (*Fig. 24*). In cyclohexanes this stereoelectronic requirement is met when the groups are diaxial but not when they are diequatorial or axial-equatorial. Accordingly, 2β,3α-dibromo-5α-cholestane *196* reacts with potassium iodide in acetone to give

* However, see D. C. Wigfield, *Tetrahedron*, 1979, **35,** 449 for further discussion.

5α-cholest-2-ene *197* while there is no detectable reaction for the epimeric 2α,3β-dibromo-compound *198* and similarly, 3α,4β-dibromo-5α-cholestane *199* reacts under similar conditions much more rapidly than the 3β,4α-dibromo-compound *200* to give 5α-cholest-3-ene *201*. It is estimated that the rate of debromination of *200* relative to that for *199* is approximately 1 : 90 and it seems likely that the debromination of the former does not proceed *via* the E2 mechanism.

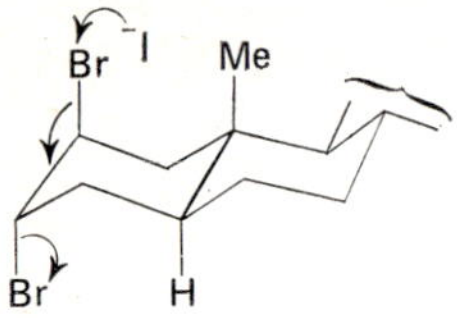

196

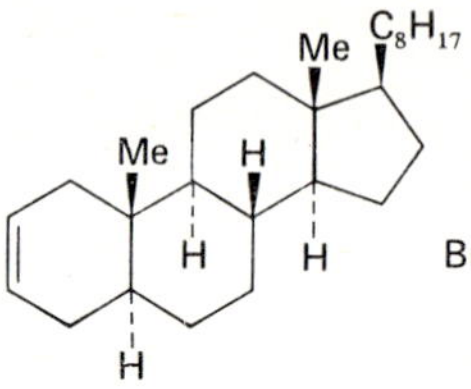

197

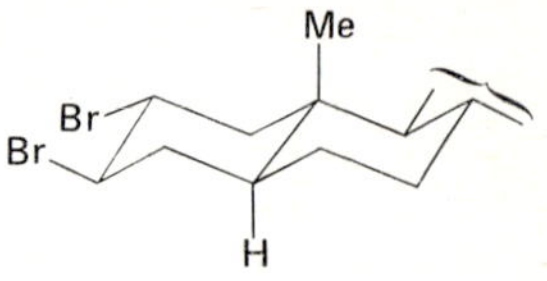

198

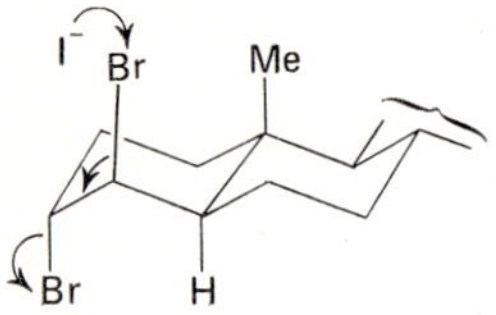

199

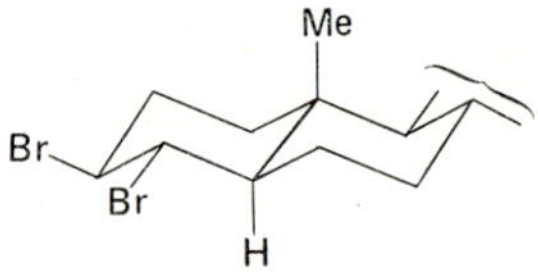

200

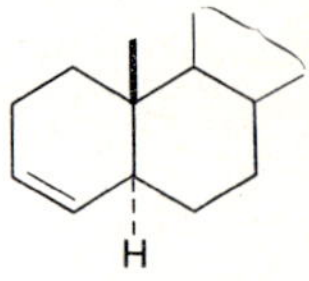

201

Sodium ethoxide-catalysed dehydrochlorination of menthyl chloride *202* to menth-2-ene *204*, presumably *via* the energetically unfavourable conformer *203*, proceeds approximately 200 times more slowly than the reaction of neomenthyl chloride *205* which in addition to *204* gives menth-3-ene *206* since axial hydrogen atoms are available at C-4 and C-2. This further demonstrates the importance of

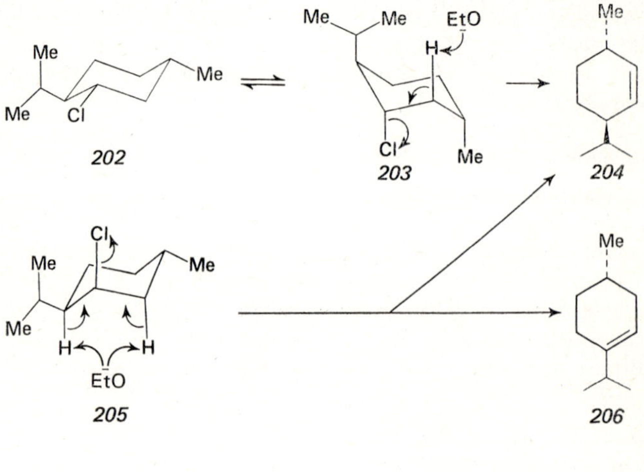

stereoelectronic factors in E2 reactions and shows also the effect on reaction rate of the ground state populations of the reacting conformers. The latter point has not been directly alluded to previously since the reactants discussed were conformationally biased or fixed. In general terms the overall rate of a reaction involving a reactant in various conformers will depend on the rate constants for the individual conformers and their populations.* The rate constants for the reactions *203*→*204* and *205*→*204* are approximately equal while that for *202*→*204* is zero. Thus *204* is produced more slowly from menthyl chloride (*202*⇌*203*) than from neomenthyl chloride *205* owing to the low ground state population of *203*. The relatively greater reaction rate (*200*) of menthyl chloride (*202*=*203*) *vs* neomenthylchloride *205* is owed in part to this but largely it is owed to the production of *206*, the major product (78 per cent).

Electrophilic addition to alkenes and nucleophilic cleavage of epoxides

The addition of halogens and hypohalous acids to cyclohexenes leads to the diaxial *trans*-product *via* a halonium ion. The stereoelectronic requirement of these reactions in general is that the nucleophile must attack the halonium ion at the side opposite from the halogen atom and that in the transition state the partial bonds are co-planar with the carbon atoms which formed the original double bond (*Fig. 27*). The same stereoelectronic factors apply in the nucleophilic opening of epoxides (*Fig. 28*) and accordingly in epoxycyclohexanes the products are diaxial.

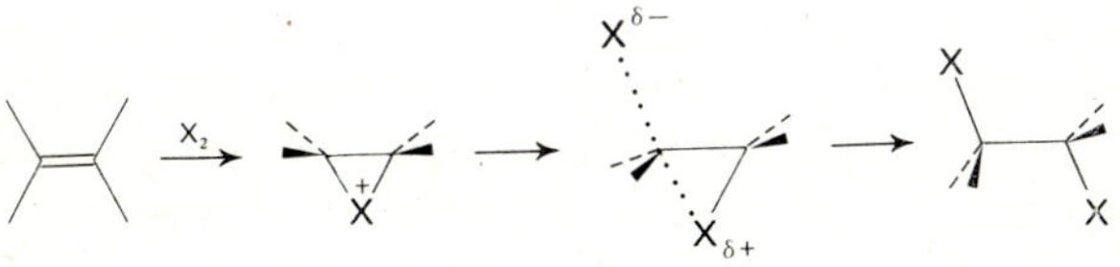

Transition state

Fig. 27. Halogen addition to cyclohexenes.

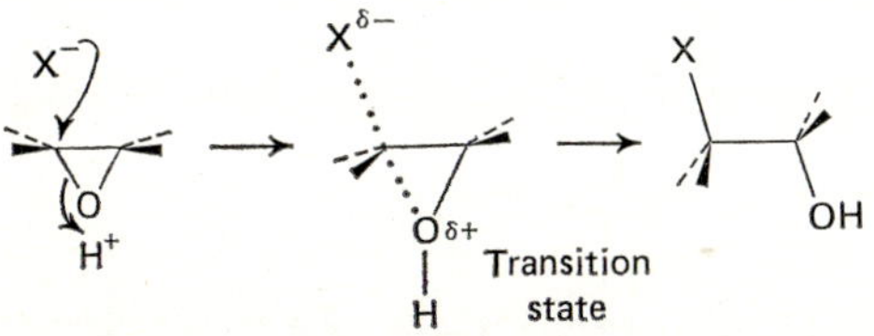

Fig. 28. Nucleophilic opening of epoxides.

* For a conformationally heterogeneous system the overall rate constant $k=\Sigma x_i k_i$ where x_i is mole fraction of the ith conformer and k_i is the rate constant for its reaction.

These reactions are exemplified by the addition of bromine to 4-*t*-butylcyclohexene *207* to give largely the diaxial dibromo-compound *208* and the cleavage of 2α,3α-epoxy-5α-cholestane *209* with hydrobromic acid to give the diaxial bromohydrin *210*. The conformation of *207* and other cyclohexenes is a half-chair in which C-1, C-2, C-3, and C-6 are co-planar and C-4 or C-5 are respectively above and below the plane. The conformations of the bromonium ions intermediate between *207* and *208* are similar to this as is the conformation of ring A in the epoxide *209*.

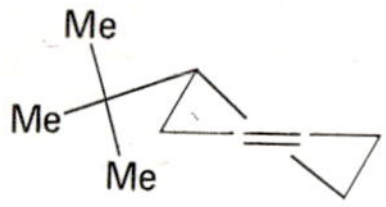

207

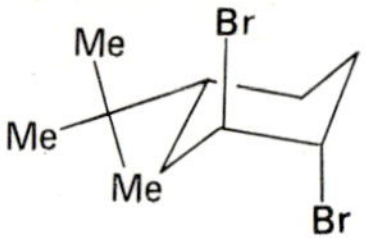

208

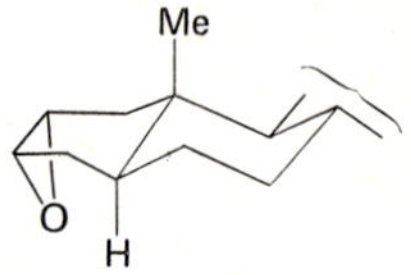

209

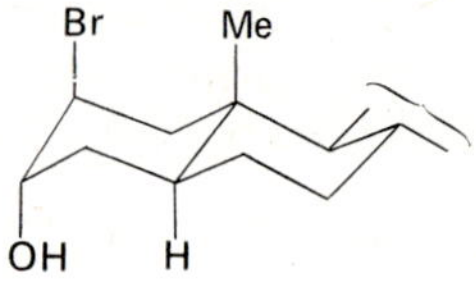

210

Reaction of 5α-cholest-2-ene *197* with hypobromous acid gives predominantly the 2β-hydroxy-3α-bromo-compound *212* *via* the 2α,3α-bromonium ion *211* which is formed preferentially owing to the steric hindrance of the 10β-methyl group. As expected from *Fig. 28* the rate of conversion of *212* in base to the 2β,3β-epoxide *213* is relatively much greater (10^4) than that for the 2α-bromo-3β-hydroxy-isomer *214*

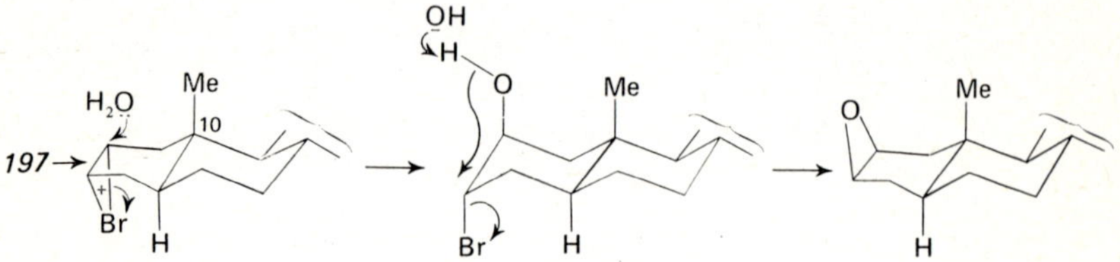

211 *212* *213*

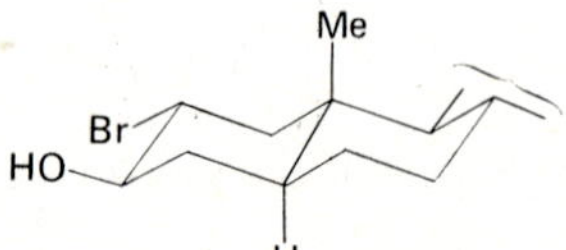

214

in which the hydroxy group and the bromine atom are equatorial. The diaxial arrangement of the reacting groups is clearly as necessary for the epoxide formation reaction as it is for the epoxide cleavage reaction.

S_N2 reactions

The rate determining step of these bimolecular nucleophilic substitution reactions involves the synchronous attack of the nucleophile and departure of the leaving groups. A stereoelectronic requirement is that the transition state has a linear arrangement of the partially bonded nucleophile, the carbon atom at which substitution is taking place, and the partially bonded leaving group (*Fig. 29*). Such sub-

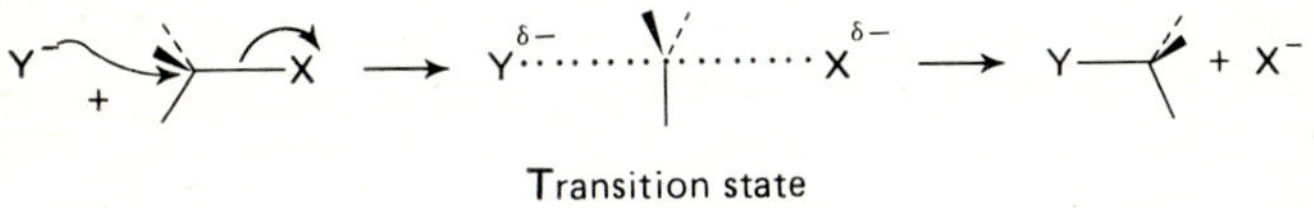

Fig. 29. S_N2 reaction.

stitutions in cyclohexanes proceed more rapidly for axial than for equatorial derivatives. For example, in reactions with sodium thiophenoxide in 87 per cent aqueous ethanol, *cis*-4-*t*-butylcyclohexyl bromide *215* reacts approximately 60 times more quickly than the *trans*-isomer *216* and the *cis*-tosylate *188* reacts approximately 30 times more quickly than the *trans*-tosylate *189*. The *cis*-derivatives *215* and *188* have higher ground state energies than the corresponding

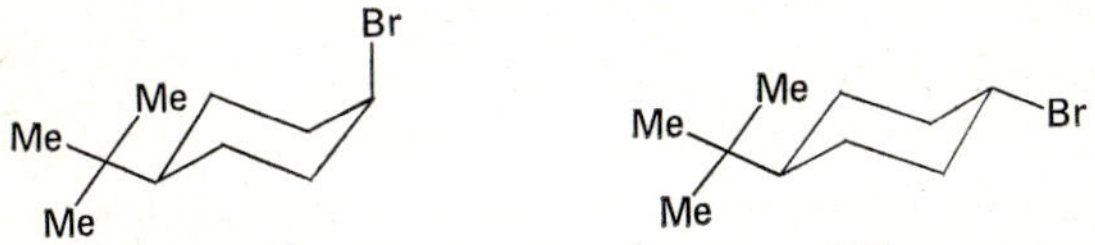

215 *216*

trans-derivatives and the transition states derived from the *trans*-derivatives are of higher energy than those from the corresponding *cis*-derivatives since the axial approach of the nucleophile (PhS^-) is more sterically hindered than the equatorial approach (*Fig. 30*). The interaction of the leaving group in the transition state *217* appears

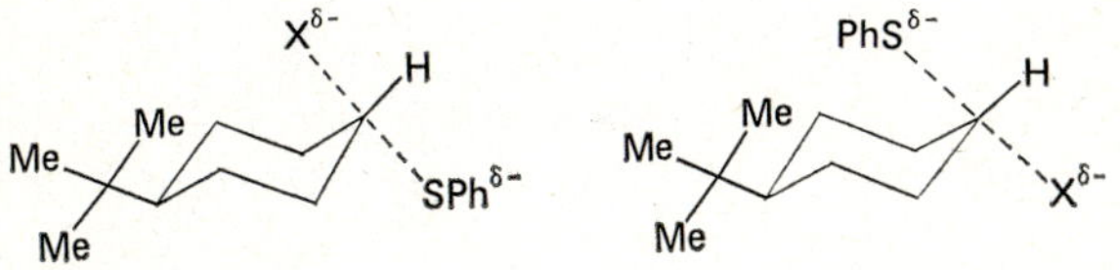

217 *218*

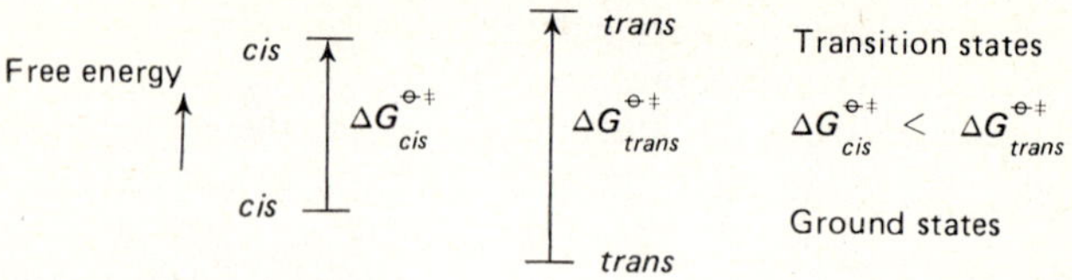

Fig. 30. Comparative activation energies for S_N2 reactions of *cis*- and *trans*-4-*t*-butylcyclohexyl derivatives.

to be much less important than the similar interactions with the approaching nucleophile in the transition state *218*.

1,2-rearrangements of carbocations

The formation of a carbocation may lead to the rearrangement of a group attached to a neighbouring carbon atom as indicated in *Fig. 31*. Stereoelectronic factors are important and in many cases the group which preferentially rearranges has the *anti* conformation with regard to the leaving group X. The implication of this is either that the

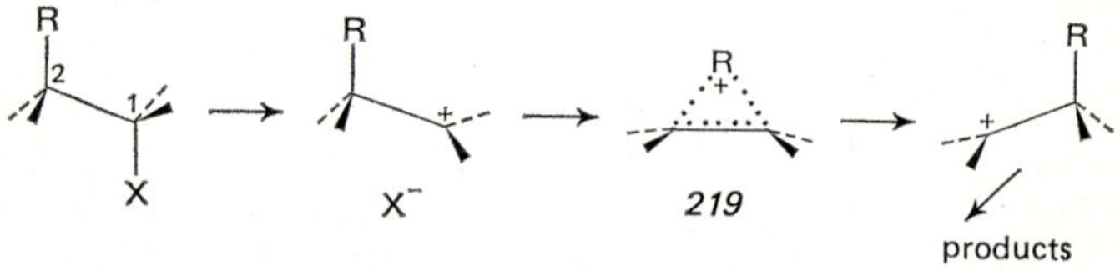

Fig. 31. Group rearrangement *via* carbocation.

migrating group R begins to move before complete ionisation of the C—X bond or that the first-formed carbocation exists as an intimate ion pair with X^-.

Thus, 3β,6β-diacetoxy-5-hydroxy-5α-cholestane *220* rearranges in acetic anhydride containing potassium bisulphate to give the alkene *222 via* the carbocation *221* whereas the 5β-epimer *223* gives the unrearranged alkene *224*. The methyl group in *220* which undergoes a 1,2-shift is *anti* to the original 5α-hydroxyl group. In *223* the methyl group and the leaving group have a *gauche* relationship and the alternative elimination is preferred as there is an *anti* hydrogen at C-4.

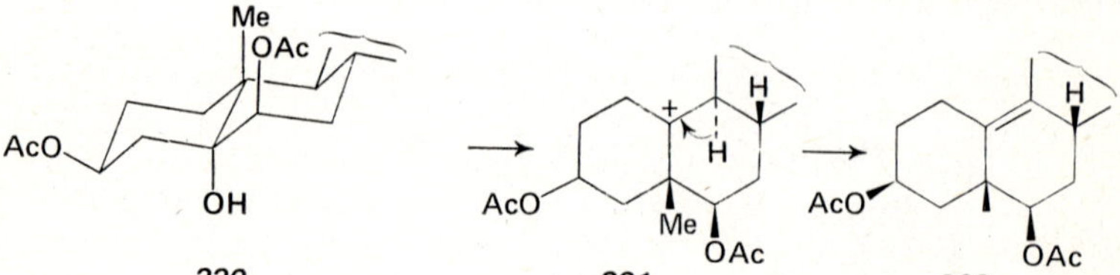

220 *221* *222*

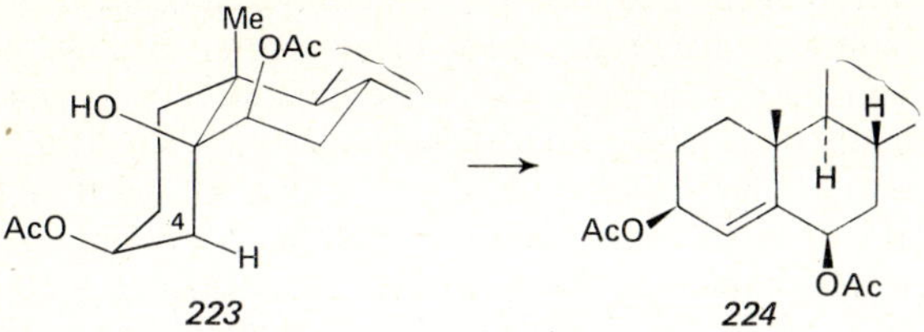

223 224

Similarly, the solvolysis of 4,4-dimethyl-5α-cholestan-3β-yl tosylate *225* gives the ring-contracted product *227* whereas the 3α-epimer *228* gives the rearranged alkene *230*. In *225* the 4,5-bond which has the *anti* conformation to the tosyloxy group migrates to give the carbocation *226*. In *228* the 4β-methyl group migrates as it has the *anti* conformation with respect to the leaving group and the intermediate carbocation *229* is formed.

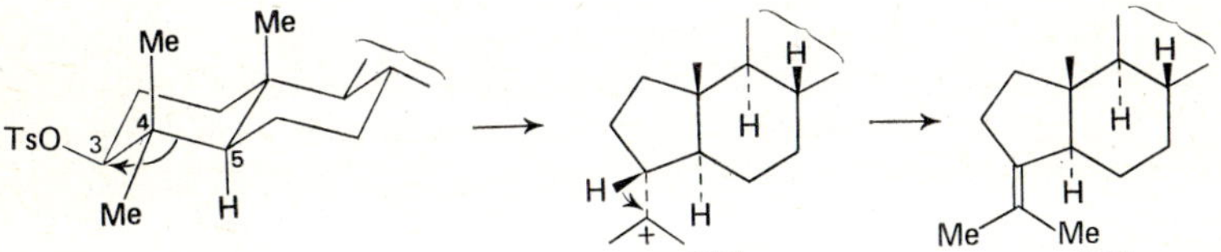

225 226 227

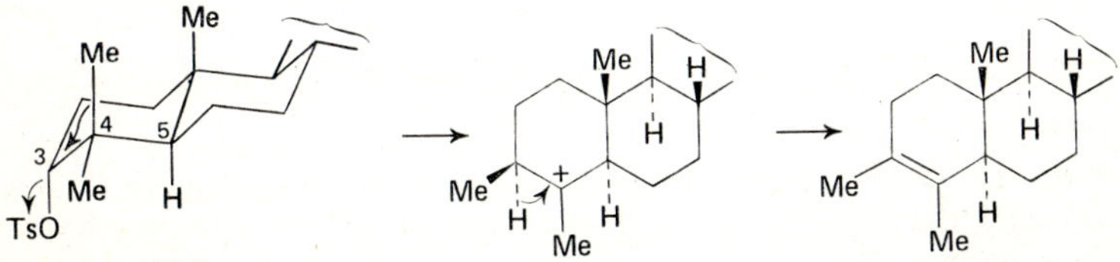

228 229 230

Comparison of reactivities of cyclopropanes, cyclobutanes, cyclopentanes and cyclohexanes

The strain present in cycloalkanes varies with the ring size (*Chapter 2*), three- and four-membered rings being particularly strained and the five-membered ring being slightly strained relative to cyclohexane. An indication of the strain present is given by the heats of combustion *per* CH_2, which taking cyclohexane = 0 are *ca* 39, 27, and 5.5 kJ mol^{-1} respectively for cyclopropane, cyclobutane, and cyclopentane. The high strain in cyclopropane and cyclobutane is largely a result of angle strain while that in cyclopentane results mainly from torsional strain.

Reactions which allow a release in the transition state of the strain present would be expected to proceed more readily than those which either preserve or increase the strain.* In the case of cyclopropanes and cyclobutanes, reactions which involve ring opening and those which involve a change from sp^2 hybridised to sp^3 hybridised ring carbon readily occur since a release of angle strain would be expected in both cases. The change from sp^2 to sp^3 may be seen to reduce the angle strain since the associated reduction in bond angles (120° to 109° 28′) means that the distortion of the formal C—C—C bond angles will be less. This simple approach is useful although, as was indicated in Chapter 2, angle strain is more accurately described by poor orbital overlap and the hybridisation in these small rings is not precisely sp^2 or sp^3. In the cyclopentanes and cyclohexanes changes in torsional strain are of particular importance. A change of hybridisation of a ring carbon from sp^2 to sp^3 is favourable for cyclohexanes, but unfavourable for cyclopentanes. In cyclohexanes, in the chair conformation, all vicinal bonds are staggered when all the ring carbon atoms are sp^3 hybridised. When one carbon is sp^2 hybridised, eclipsing occurs between the bond joining the substituent atom to the sp^2 hybridised carbon and the vicinal bonds. Thus in cyclohexanone *231* the C=O is eclipsed by the vicinal equatorial C—H bonds. In cyclopentanes there is considerably less eclipsing

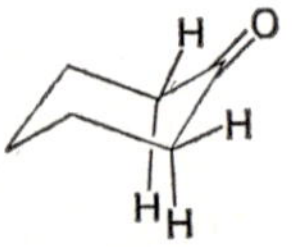

231

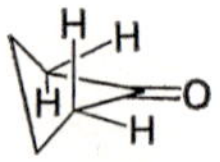

232

when one carbon is sp^2 hybridised. In cyclopentanone *232*,† for example, the C=O has a staggered conformation with respect to the vicinal C—H bonds.

The following reactions illustrate the importance of the effects of strain on reactivity.

Ring opening reactions

It would be expected from the foregoing discussion that cyclopropanes and cyclobutanes would undergo ring opening rather more easily than cyclopentanes and cyclohexanes. This is confirmed by hydrogenolysis in the presence of a nickel catalyst where the temperature required to convert the cycloalkane to its corresponding alkane

* The concept that reactivity could be correlated with ring strain changes was first proposed by H. C. Brown who used the term I-strain to describe the internal strain present.

† This is shown in the half-chair conformation rather than the envelope (*see 105, Chapter 2*) as there is considerable evidence to suggest that the former is considerably more stable than the latter. The equivalent conformations in cyclopentane do not differ significantly in energy.

is respectively 80°C, 180°C and 300°C for cyclopropane, cyclobutane and cyclopentane.

A number of electrophilic reagents react readily with cyclopropane but not with the other cycloalkanes. Thus bromine reacts with cyclopropane to give 1,3-dibromopropane. The failure of the other cycloalkanes to undergo this type of ring opening relates to their lower strain energies and their failure to achieve a transition state in which the strain energy is partially released.

Nucleophilic additions to carbonyl groups

The rates of the sodium borohydride reductions of cyclobutanone, cyclopentanone and cyclohexanone to the corresponding alcohols (*Fig. 32*) at 0 °C are in the ratio *ca* 38:1:23. These reactions, which involve a change of hybridisation of the ring carbon from sp^2 to sp^3, proceed much more quickly for cyclobutanone and cyclohexanone owing to the relief of strain associated with these changes.

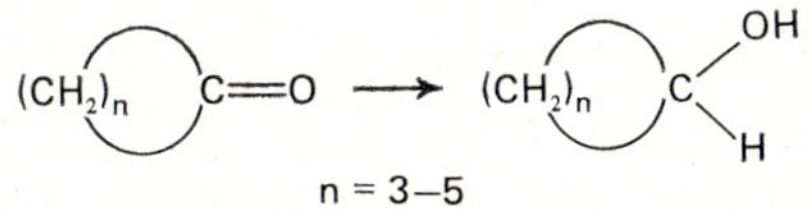

Fig. 32. Hydride reduction of cycloalkanones.

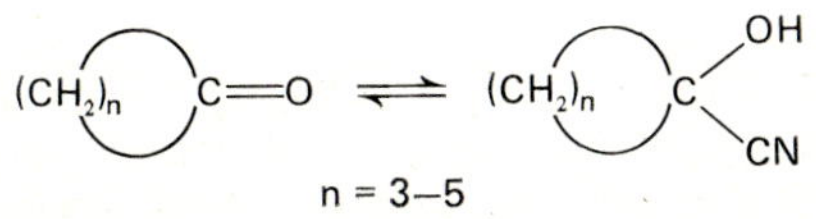

Fig. 33. Hydrogen cyanide addition to cycloalkanones.

A similar though not identical trend is obtained for the equilibrium constants for the addition of hydrogen cyanide to cyclobutanone, cyclopentanone and cyclohexanone (*Fig. 33*) which are in the ratio *ca* 3:1:20. The values recorded in the literature are quite variable but, in general, that for cyclobutanone is intermediate between the other two and cyclopentanone is the least reactive as expected.

Nucleophilic substitutions

In S_N1 reactions the formation of the intermediate carbocation involves a change in hybridisation from sp^3 to sp^2 (*Fig. 25*) and in S_N2 reactions the transition state contains a pentacoordinate carbon (*Fig. 29*) which is also sp^2 hybridised. The reverse trend of reactivity of that in nucleophilic additions above would be predicted for the cycloalkanes in these reactions. This is confirmed by the acetolyses of cyclopropyl-, cyclobutyl-, cyclopentyl- and cyclohexyl-tosylates which have the relative rates 10^{-5}, 11.3, 14, and 1 respectively at 25 °C. The rate of reaction for cyclobutyl tosylate is anomalously high owing to the formation of a highly stabilised, possibly non-classical*, carbocation rather than a simple cyclobutyl cation.

* There is some controversy over the precise structure of this stabilised ion and the factors responsible for its stabilisation; see H. C. Brown, *Tetrahedron*, 1976, **32,** 179 for further discussion.

5. Conformation and Reactivity in Acyclic Systems

The high conformational mobility of acyclic molecules contrasts with that of the cyclohexanes which with appropriate substitution may be conformationally biased or fixed (*Chapters 2 and 4*). Correlations of conformation and reactivity are thus less accurate for acyclic molecules. Despite this, meaningful correlations can be made and steric and stereoelectronic factors are shown to be important. Their influence is illustrated by the following reactions.

Electrophilic addition to alkenes and E2 eliminations

Addition of halogens and hypohalous acids to alkenes usually proceeds in a stereospecifically *trans* manner (*see Chapter 4*). Accordingly, *cis*-but-2-ene *3* and *trans*-but-2-ene *4* react with bromine *via* the bromonium ions *233* and *236* to give racemic 2,3-dibromobutane *234* and *meso*-2,3-dibromobutane *237* respectively (enantiomers of *234* and *236* are omitted*). The diastereoisomers *234* and *237* undergo E2 elimination with potassium iodide in acetone *via* the transition states *235* and *238* to give respectively *cis*-but-2-ene *3* and *trans*-but-2-ene *4* owing to the stereoelectronic requirement that the two bromine atoms have the anti-conformation (*see Chapter 4*).

The rates of elimination of *234* and *237* are in the ratio 1 : 2. In the transition states *235* and *238* there is some flattening of the carbon atoms which are partially bonded to the bromine atoms and the groups in a *cis*-relationship move towards an eclipsed arrangement.

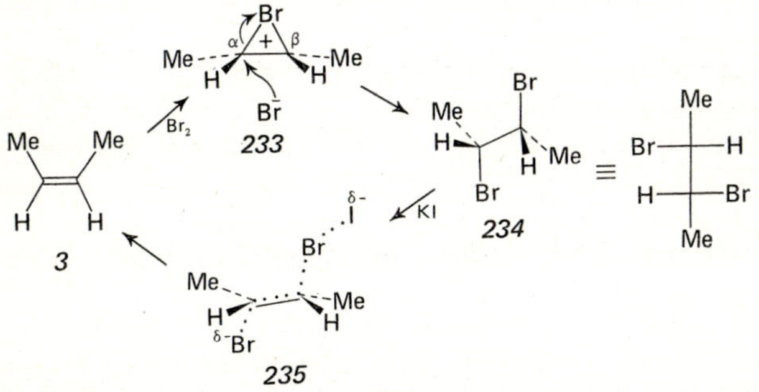

* The enantiomer of *234* would arise from attack of Br^- on the alternative β-carbon in *233* and the enantiomer of *236* would arise from attack of Br_2 on the alternative face of *4*. Irrespective of which enantiomer of *236* is involved and irrespective of which of its α- or β-carbon atoms is attacked by Br^-, the product is the *meso*-compound *237*.

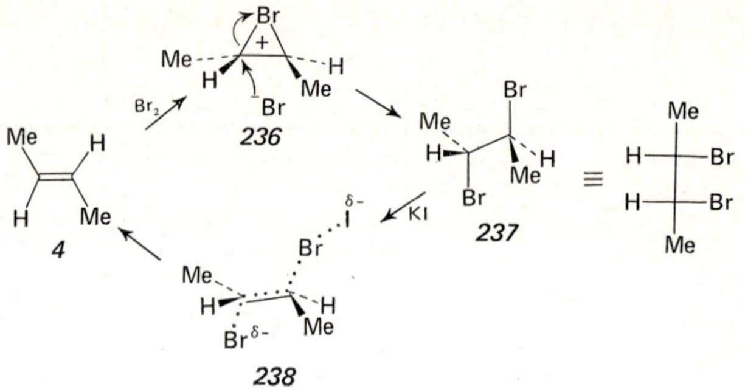

In *235* the eclipsing occurs between two methyl groups while in *238* it occurs between a methyl group and a hydrogen atom. It might be anticipated that the transition state *235* would thus have significantly higher energy than *238* and accordingly the reaction *via* the former from racemic 2,3-dibromobutane *234* would be faster than the reaction *via* the latter from *meso*-2,3-dibromobutane. This rationalisation is possibly facile, however, since it ignores the ground state conformational preferences of the diastereoisomers and assumes that *234* and *237* do not differ significantly in energy. These factors may be very significant in this case and others where the relative rates of reaction do not differ greatly. It has been suggested* that in this particular case the difference in reaction rates is largely owed to the favourable ground state population (*see Chapter 4*) of *237* *vs* *234* and that the rate constants for the two reactions are the same owing to the retention of much the same steric interaction in the transition state *235* as is present in the reacting conformer *234*. The eclipsing effect of the *cis*-methyls is believed to be sufficiently ameliorated by their movement apart as the carbon atoms, to which they are bonded, become sp^2 hybridised.

In the more general situation, the eclipsing of *cis*-groups in E2 reactions has a profound effect and considerable differences in reaction rates of diastereoisomers may be observed. In these cases, small differences in the energies of the ground states are barely significant and the influence of the ground state population of the reacting conformer is relatively small. For example, racemic stilbene dichloride *239* (only one enantiomer shown) gives, on heating with pyridine at 200 °C, the *trans*-chlorostilbene *240* whereas under the same conditions, *meso*-stilbene dichloride *241* fails to give the expected *cis*-chlorostilbene *242*. The eclipsing of the phenyl groups in the transition state leading to *242* is so severe that the transition state reaches an energy which is unattainable under these conditions.

* E. L. Eliel, N. L. Allinger, S. J. Angyal, and G. A. Morrison, *Conformational analysis*. New York: Wiley Interscience, 1966, p 28.

In the above examples only one conformer of the substrate can react. There are cases where two (or more) conformers of a substrate may react to give two (or more) different products. It has been shown that, in general, the overall reaction rate of a conformationally heterogeneous substrate depends on the ground state populations of the reacting conformers. However, when the conformers may give rise to different products, the product ratio does not depend on these populations, but is determined by the relative standard free energy

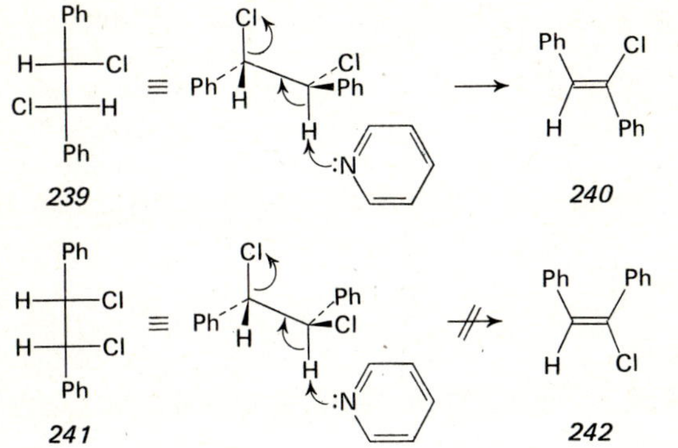

levels of the transition states leading to the various products, provided the standard free energies of activation for each of the reactions is large compared with the standard free energy barriers to interconversion of conformers. This is the **Curtin-Hammett principle** and it is quite general and may be derived mathematically from the transition state theory. It may be exemplified by consideration of the base-catalysed dehydrochlorination (E2 elimination) of 1,2-diphenyl-1-chloroethane *243* (one enantiomer is shown) to *cis*-stilbene *246* and *trans*-stilbene *247*. The conformers *244* and *245*

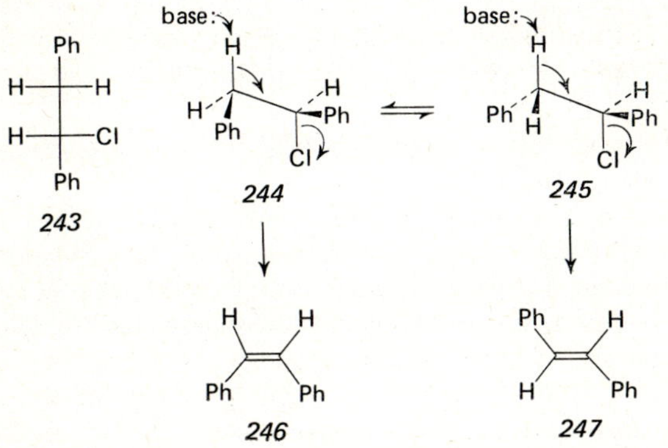

have the required *anti*-conformation of the H and Cl and the former gives rise to *cis*-stilbene *246* and the latter gives rise to the *trans*-isomer *247*. That the *trans*-isomer *247* is the major product does not relate to the position of the equilibrium between *244* and *245*. Rather, it is because *244* leads to a transition state in which the *cis*-phenyl groups become eclipsed and which therefore has a much higher energy than the transition state from *245* in which a phenyl group and a hydrogen atom become eclipsed.

An important consequence of the Curtin-Hammett principle is that a major product may be derived from an unfavourable ground state conformer provided the transition state energy is favourable. This situation is depicted in *Fig. 34* for a substrate existing in two conformers A and B. Provided $\Delta G_A^{\ominus\ddagger}$ and $\Delta G_B^{\ominus\ddagger}$ are large compared with $\Delta G_{A\rightarrow B}^{\ominus\ddagger}$ the relative energies of A and B are unimportant and the major product will be derived from the lower energy transition state.

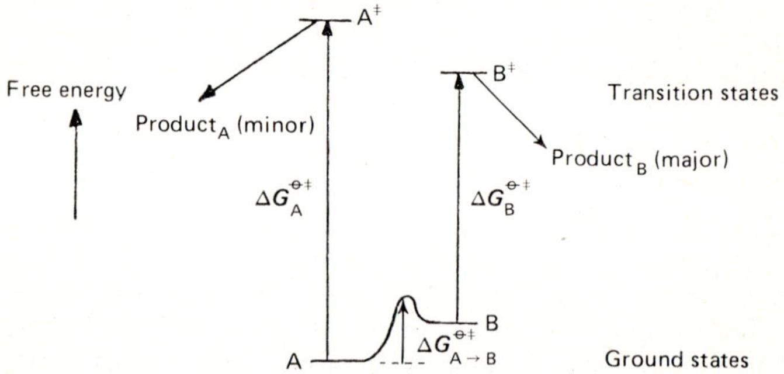

Fig. 34. Curtin-Hammett principle.

It should be stressed that the Curtin-Hammett principle applies to conformationally heterogeneous substrates in which the various conformers present may give rise to different products. It has no relevance to the study of the comparative reactions of diastereoisomers such as *234* and *237* which, of course, are different substrates and not conformational isomers.

Asymmetric induction

In general terms, the introduction of a new chiral centre in a molecule already containing a chiral centre would be expected to give an unequal mixture of diastereoisomers since the transition states leading to the two products would be diastereomeric and therefore have different energies. This asymmetric induction is commonly observed with carbonyl addition reactions where the faces of the carbonyl group are diastereotopic (*see Chapter 1*).

In the case of kinetically controlled addition reactions of carbonyl compounds (*eg* reactions with Grignard reagents and hydride reducing agents) of the type LMSCCOR, in which L, M and S represent the large, medium and small groups respectively, Cram's rule predicts the stereochemical outcome. The ketone may be regarded as reacting in conformer *248* with the reagent (XY) approaching from the same side as the small group S giving rise to the product *249*.

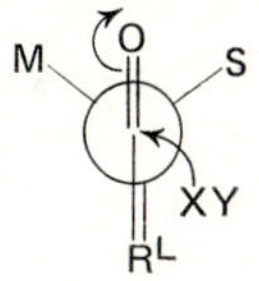

248

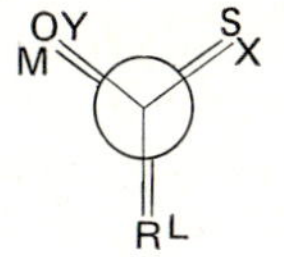

249

Thus lithium aluminium hydride reduction of benzoin *250* (one enantiomer shown) gives largely *meso*-hydrobenzoin *251*.

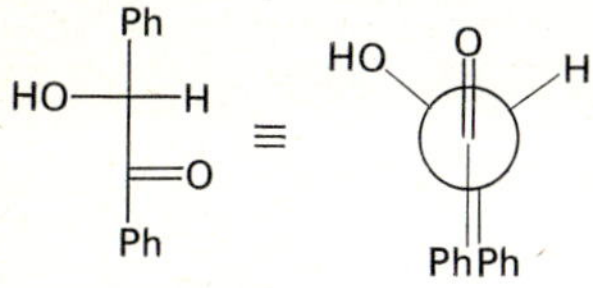

250

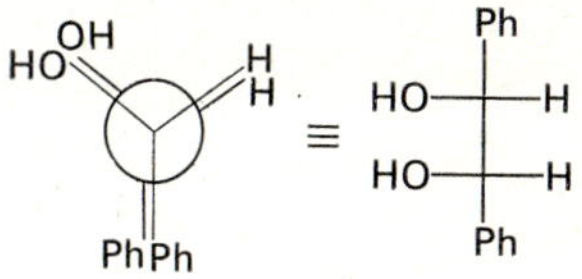

251

Prelog's rule predicts the stereochemical outcome of similar reactions with α-keto esters of the type RCOCOOCSML. The ketone may be regarded as reacting in the conformer *252* in which the two carbonyl groups are *anti* and the group S is in the same plane as and *syn* to the ketone carbonyl group. The reagent (XY) approaches from the same side as the smaller (M) of the two remaining groups, giving rise to the product *253*.

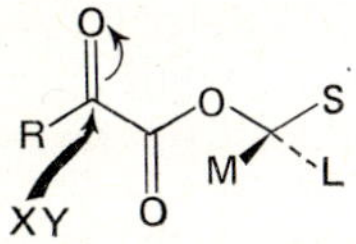

252

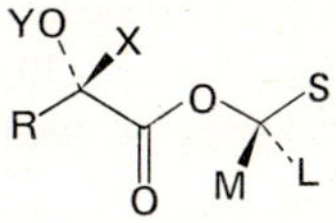

253

Thus, (−)-menthyl phenylglyoxylate *254* reacts with methyl magnesium iodide to give mainly the derivative *255* which on hydrolysis gives (−)-atrolactic acid *256*.

It should not be inferred that *248* and *252* necessarily have any mechanistic significance. The Cram and Prelog rules are based on

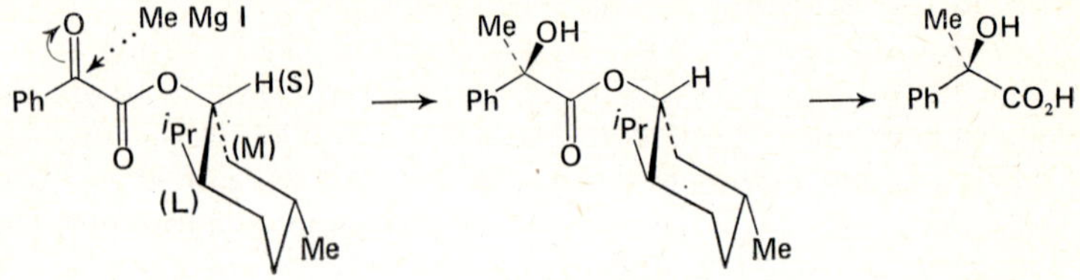

254 *255* *256*

empirical observations and it is unlikely that the transition states of the reactions result from *248* and *252* which are merely convenient models allowing reasonable predictions to be made. Indeed, Felkin and his co-workers have proposed that the transition states for the reactions of the compounds LMSCCOR are reactant-like and, in the absence of polar groups, are best represented as *257* to minimise steric and torsional strains. A major advantage of this view* is that, unlike the Cram rule, it readily accounts for the observed stereoselectivity both in acyclic ketones and cyclohexanones (*see Chapter 4*).

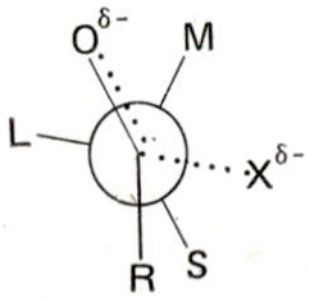

257

1,2-rearrangements and neighbouring group effects

As indicated in *Fig. 31* (*Chapter 4*) the preferred migrating group in 1,2-rearrangements of carbocations is often that which has the *anti* conformation with regard to the leaving group. At some stage the migrating group becomes bonded to both carbon atoms 1 and 2 and the stability of this bridged ion *219* depends on the nature of the migrating group. In certain cases the bridged ion may be sufficiently stable and have a long enough life time to be a true intermediate rather than a transition state. The formation of such intermediate ions by participation of neighbouring groups can have profound stereochemical and kinetic effects.

For example, the reactions of racemic *erythro*-3-bromobutan-2-ol *258* and racemic *threo*-3-bromobutan-2-ol *259* with hydrobromic acid give respectively *meso*-2,3-dibromobutane *237* and racemic 2,3-dibromobutane *234*. Participation by the neighbouring *anti* bromine atom in *258* and *259* gives the intermediate bromonium ions

* M. Cherest, H. Felkin, and N. Prudent, *Tetrahedron Letters*, 1968, 2199.

236 and *233* respectively which are attacked by bromide ion as indicated. Only one of the possible enantiomers is shown for *258*, *259*, *236* and *234* and only one mode of opening of the bromonium ions *233* and *236* is indicated although attack by bromide ion at the α- and β-carbon atoms is equally likely (see earlier). The proposed mechanism is further supported by the observation that a single

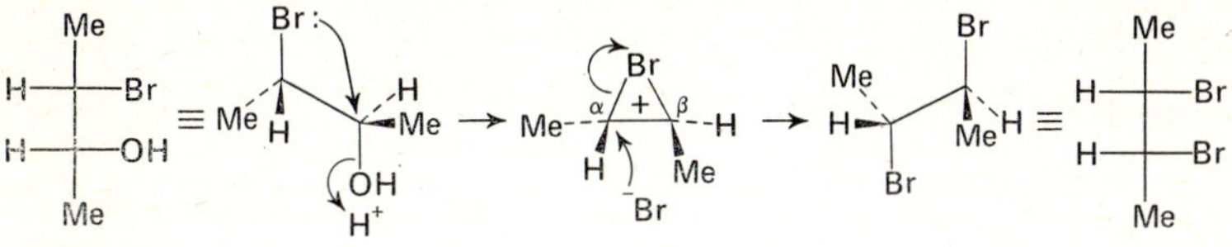

258 *236* *237*

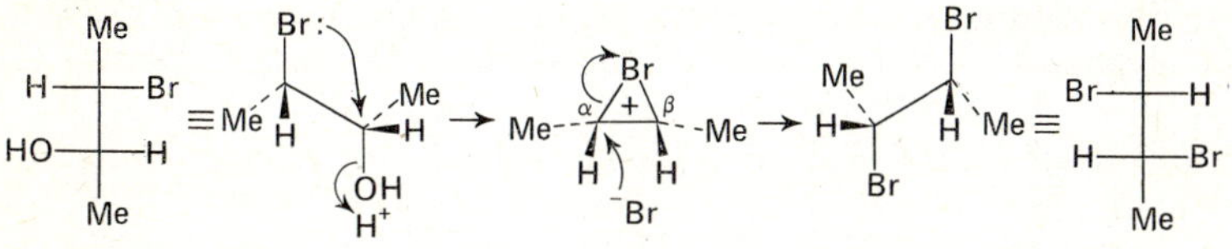

259 *233* *234*

enantiomer of *threo*-3-bromobutan-2-ol *259* gives racemic 2,3-dibromobutane *234*. The important observation to be made in these reactions is that the relative configurations of the two chiral centres remain unchanged and no other reaction mechanism could satisfactorily account for this.

Similar stereochemical effects are observed in the reactions of racemic *threo*-2-bromo-3-methoxybutane *260* and the racemic *erythro*-isomer *263* with silver acetate in acetic acid. The former gives racemic *threo*-2-acetoxy-3-methoxybutane *262 via* the bridged oxonium ion *261* and the latter gives racemic *erythro*-2-acetoxy-3-methoxybutane *265 via* the similar bridged oxonium ion *264*. Only a single enantiomer is shown for *260*, *262*, *263*, *264* and *265* and only one mode of opening of the oxonium ions *261* and *264* is indicated. It is interesting to note that, whereas the alternative attack of acetate ion at the β-carbon in *261* gives the enantiomer of *262* in *264* attack at the

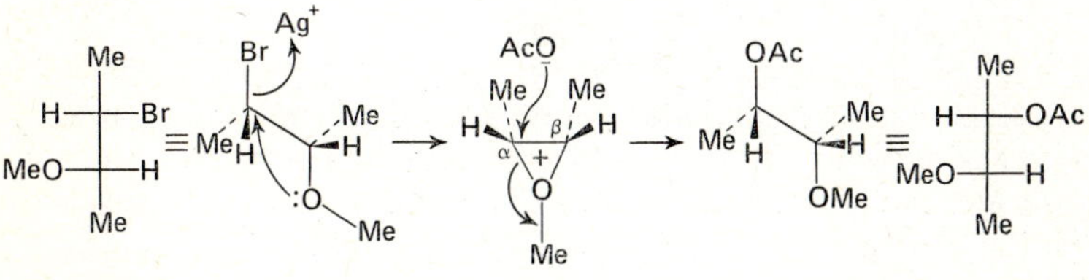

260 *261* *262*

α- or β-carbon gives a single enantiomer *265*. This relates to the fact that *264* is chiral whereas *261* is not. Although *236* is similarly chiral it, of course, cannot give a single enantiomer on opening with bromide ion since the product is the *meso*-compound *237*.

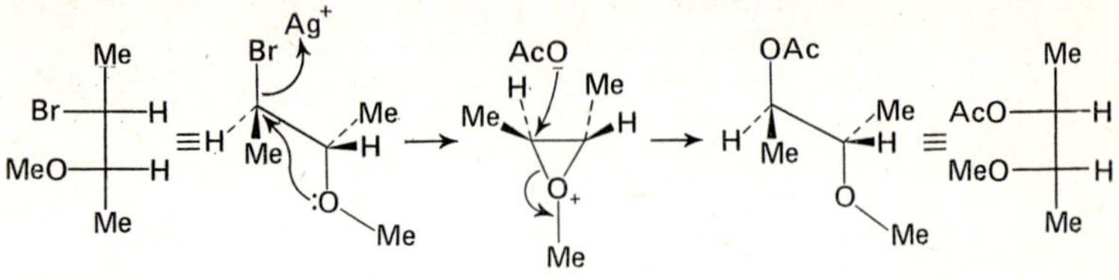

263 264 265

In addition to participation by neighbouring groups which have a lone pair of electrons and therefore may obviously behave in a nucleophilic manner, it is possible for substituted phenyl groups to behave similarly. Thus acetolysis of optically active *threo*-3-phenylbutan-2-yl tosylate *266* gives racemic* *threo*-3-phenylbutan-2-yl acetate *268 via* the phenonium ion *267* which is achiral, while optically active *erythro*-3-phenylbutan-2-yl tosylate *269* gives optically active *erythro*-3-phenylbutan-2-yl acetate *271 via* the chiral phenonium ion *270*.

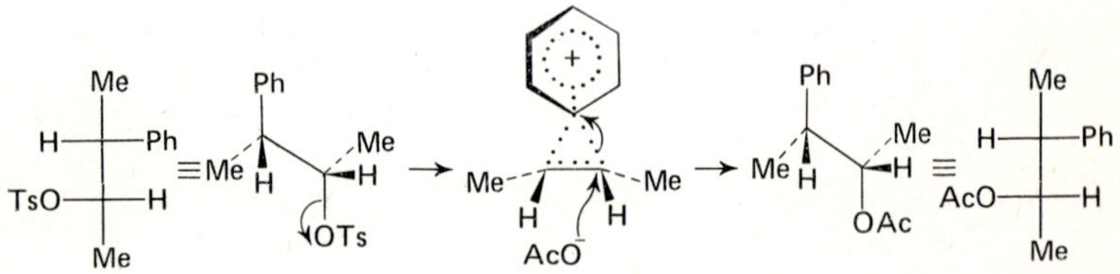

266 267 268

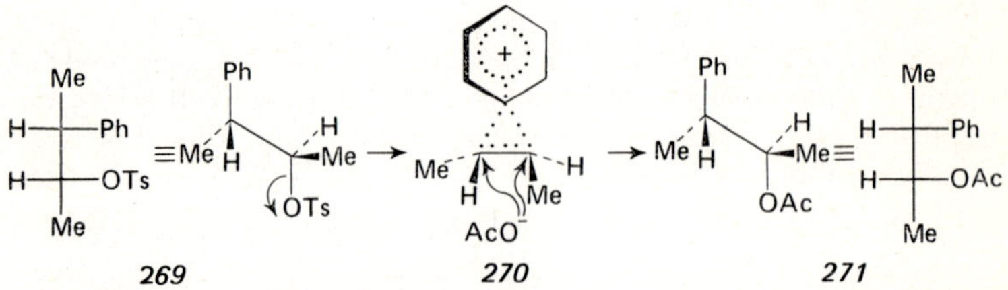

269 270 271

Neighbouring group participation and 1,2-rearrangements usually occur with suitable groups which are *anti* to the leaving group and, as indicated in *Chapter 4*, the implication of this may be that the participation or migration is concerted with ionisation. This is shown to

* One enantiomer is omitted.

be so for *266* and *269* by virtue of a considerable rate enhancement observed for 3-(*p*-methoxyphenyl)-butan-2-yl tosylate *272*. Electron release from the *p*-methoxy group would be expected to stabilise the intermediate phenonium ion and the transition state leading to it.

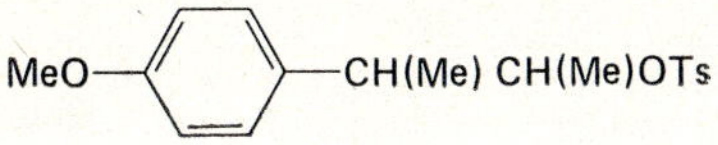

272

In the examples considered in this section no formal rearrangement of groups is apparent although the formation of bridged intermediate ions is evident from stereochemical and kinetic data. In the acetolysis of racemic *threo*-2-phenylpentan-3-yl tosylate *273** the isolation of the rearranged product, racemic *threo*-3-phenylpentan-2-yl acetate *275** in addition to racemic *threo*-2-phenylpentan-3-yl acetate *276** confirms the intermediacy of the phenonium ion *274*.*

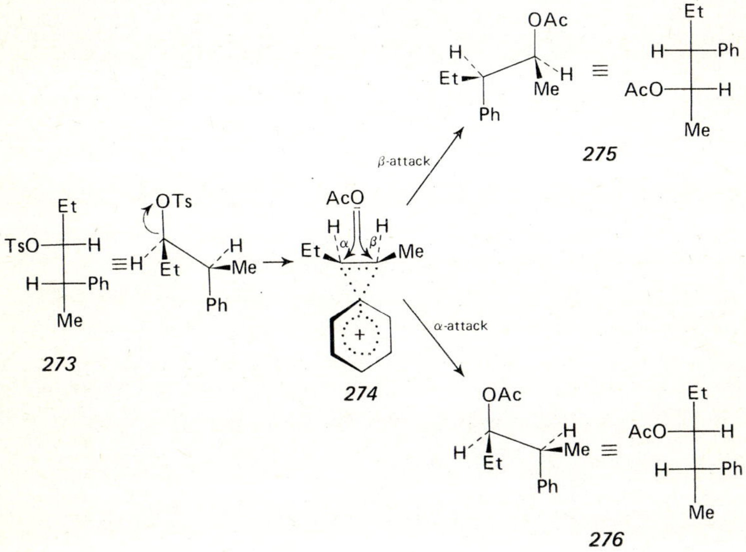

* One enantiomer is omitted.

Suggestions for Further Reading

E. L. Eliel, *Elements of stereochemistry*. New York: Wiley, 1969.

G. Natta and M. Farina, *Stereochemistry*. London: Longman, 1972.

B. Testa, *Principles of organic stereochemistry*. New York: Dekker, 1979.

H. Kagan, *Organic stereochemistry*. London: Arnold, 1979.

E. L. Eliel, *Stereochemistry of carbon compounds*. New York: McGraw-Hill, 1962.

E. L. Eliel, N. L. Allinger, S. J. Angyal, and G. A. Morrison, *Conformational analysis*. New York: Wiley-Interscience, 1965.

M. Hanack, *Conformation theory*. Academic Press: London, 1965.

G. H. Whitham, *Alicyclic chemistry*. London: Oldbourne, 1963.

F. J. McQuillin, *Alicyclic chemistry*. London: Cambridge, 1972.

N. G. Clark, *The shapes of organic molecules*. London: Murray, 1977.

K. Bláha, O. Červinka, and J. Kovář, *The fundamentals of stereochemistry and conformational analysis*. London: Butterworth, 1971.